ADVANCED GENERAL STATISTICS

by the same author

ELEMENTARY STATISTICS

Advanced General Statistics

B. C. ERRICKER, B.Sc.
Formerly Senior Mathematics Master
County High School, Leyton

HODDER AND STOUGHTON

LONDON SYDNEY AUCKLAND TORONTO

ISBN 0 340 15178 1

First printed 1971
Reprinted (with corrections) 1972, 1973, 1975, 1976, 1977, 1978, 1979

Printed in Great Britain for
Hodder and Stoughton Educational,
a division of Hodder and Stoughton Ltd.,
Mill Road, Dunton Green, Sevenoaks, Kent TN13 2YD,
by William Clowes & Sons Limited
Beccles and London

Preface

Many students who visit me after they leave school and have entered the universities or other institutions of learning ask me if I know of a book on statistics that they could read on their own and would give them a good working knowledge of the subject. They are often surprised to find that in their further studies a working knowledge of the methods of statistics is an essential requirement. Their knowledge of mathematics varies from 'O' to 'A' level of the General Certificate of Education and sometimes beyond this. Thus many of these individuals cannot read books on the mathematical theory of statistics while others can—up to a point. As far as I know there is no book that adequately satisfies the requirements of these students, and so I determined to experiment and try to produce one. This book is the result, and I hope it will fulfil its purpose.

Those students with no more knowledge of mathematics than is necessary for the 'O' level of the G.C.E. will be able to read the text, and since there are many worked illustrative examples they should be able to obtain a good working knowledge of statistics. The only mathematics with which they are likely to be unfamiliar is Permutations and Combinations and the Functional and Σ Notation. These are simply and adequately explained in the book, and their understanding should not be beyond these students' comprehension. Avoidance of mathematics is a necessity with some students who often have great abilities in other directions. Some of these may find one or two 'sticky patches' but they can ignore them and accept the results. One of the difficulties of statistics is the large amount of notation needed and the absence of a standard notation. In this book an attempt has been made to reduce the amount of notation to a minimum and to keep the symbols used as simple and familiar as possible.

Thus, I hope, those students who study Biology, Economics, Sociology,

Education, Medicine and kindred subjects and who find that they have to attend 'crash' courses on statistics in their studies will be able to read this book and derive pleasure from it. The explanations are detailed and the worked examples very varied.

Those pupils requiring a knowledge of statistics for 'A' level examination purposes will find the relevant mathematics at the end of each chapter. The book more than adequately covers the requirements of all 'A' level examinations and many questions from the various Examination Boards are given in the book.

Those students who will specialize in mathematics, computation or the physical sciences and who intend to study statistics at a higher level will acquire a working knowledge of the subject before they meet the theoretical treatises on the subject, and so they will obtain a practical knowledge of statistics that will navigate them through the sea of advanced mathematics they are likely to meet in a theoretical study of the subject.

I have taught statistics in school to mixed classes of sixth-formers, and I am grateful to them for the encouragement which they have given me to write this book.

For permission to use questions set in examinations I wish to thank:

The Associated Examining Board for the General Certificate of Education, Additional and Advanced Level (AEB);

The University of Cambridge Local Examinations Syndicate, Advanced Level (C);

The University of London School Examinations Department, Additional and Advanced Level (L);

The Oxford & Cambridge Examination Board (O and C);

The Joint Matriculation Board (JMB).

I also thank the Controller of H. M. Stationery Office for permission to publish material from *The Annual Abstract of Statistics*, *The Monthly Digest of Statistics*, and *Economic Trends*. All such material has been denoted by the symbol (A).

I should like to thank my wife for her tolerance of periods of silence on my part while I was considering methods of presentation, and for her help in copying the text; also to Mr. E. J. Shannon who has given invaluable help in proofing the text and in other ways; the Managing Editor of The English Universities Press Ltd has always been helpful and shown great consideration.

Care has been taken to avoid mistakes in the text and answers, but if any are found they are my fault and I should be grateful if they are brought to my notice.   B.C.E.

Contents

Tables

CHAPTER 1

Functional and Σ Notation

Functional Notation

Any expression involving a variable, and whose value can be calculated when the value of the variable is known, is said to be a function of the variable. If x is the variable, the function is usually represented by $f(x)$, $F(x)$, or $\phi(x)$ or some similar symbol and is read as 'function of x'.

Suppose $f(x) = x^2 + 3x + 4$, then $f(2)$ denotes the value of the function when $x = 2$.

Thus
and

$$f(2) = 2^2 + 3(2) + 4 = 14$$
$$f(0\cdot5) = (0\cdot5)^2 + 3(0\cdot5) + 4 = 5\cdot75$$
$$f(0) = 0^2 + 3(0) + 4 = 4$$
$$f(-2) = (-2)^2 + 3(-2) + 4 = 2$$
$$f(x+a) = (x+a)^2 + 3(x+a) + 4 = x^2 + (2a+3)x + a^2 + 3a + 4$$

and so on.

The f of $f(x)$ must not be mistaken for the f which stands for the frequency of a class in a frequency distribution. The f of a frequency distribution stands for a number and fx means the product of f and x.

EXAMPLE

If $f(x) = \dfrac{3x+4}{2x^2-1} - 6$, calculate

(a) $f(0)$ (b) $f(3)$ (c) $f(a+1)$

SOLUTION

(a) $f(0) = \dfrac{3(0)+4}{2(0)^2-1} - 6 = -10$

1

(b) $f(3) = \dfrac{3(3)+4}{2(3)^2-1} - 6 = -5 \cdot 235$

(c) $f(a+1) = \dfrac{3(a+1)+4}{2(a+1)^2-1} - 6 = \dfrac{1-21a-12a^2}{1+4a+2a^2}$

Exercise 1.1

1 If $f(x) = x^{-2}$, calculate the value of $f(2)-f(0\cdot5)$.
2 If $f(x) = 3x^2-2$, calculate the value of $f(4)-f(2)+f(0)$.
3 If $f(x) = 3x^3+ax^2+bx+4$ and $f(1) = 8$ and $f(2) = 26$, calculate the values of a and b.
4 If $f(1) = 0$ and $f(2) = 0$, state two roots of $f(x) = 0$.
5 If $F(x) = x^3-2x$, calculate the value of $2F(1)+3F(2)$.
6 If $\phi(x) = (x^2-4)(x^2-9)$, prove that $\phi(2) = \phi(-2) = \phi(3) = \phi(-3) = 0$.
7 If $f(x) = a^x$, prove that $f(x) \times f(y)/f(z) = a^{x+y-z}$.
8 If $f(x) = \log x$, prove that $f(mn) = f(m)+f(n)$.

Extension to two or more variables

The concept of a function can be extended to two or more variables. Thus if $f(x, y) = 4+3x-2y$, the value of $f(x, y)$ when $x = 2$ and $y = 3$ is found by substituting these values in the given function. Thus if

$$
\begin{aligned}
f(x, y) &= 4+3x-2y \\
f(2, 3) &= 4+3(2)-2(3) &= 4 \\
f(3, 0) &= 4+3(3)-2(0) &= 13 \\
f(-2, -2) &= 4+3(-2)-2(-2) &= 2
\end{aligned}
$$

Exercise 1.2

If $f(x, y) = 3x+2y-6$, calculate the value of

(a) $f(2, 2)$ (b) $f(3, -1)$ (c) $f(0, 0)$ (d) $f(1, 1)$ (e) $f(0, 10)$

The Σ Notation

Σ is a capital Greek letter called sigma and is used in mathematics to denote 'the sum of'. Thus $\sum\limits_{r=1}^{r=4}$ is read as 'the sum of terms like r where r

takes the values 1 to 4'. Thus

$$\sum_{r=1}^{r=4} r = 1+2+3+4 = 10$$

$$\sum_{r=1}^{r=5} r^2 = 1^2+2^2+3^2+4^2+5^2 = 55.$$

When no confusion results we denote these sums by $\sum_1^4 r$ or $\sum_1 r$ and $\sum_1^5 r^2$ or $\sum r^2$.

EXAMPLE

Find the value of

(a) $\sum_1^3 (2r+3)$ (b) $\sum_1^n (2r+3)$

SOLUTION

(a) $\sum_1^3 (2r+3) = 2(1)+3+2(2)+3+2(3)+3 = 21$

(b) $\sum_1^n (2r+3) = 2(1)+3+2(2)+3 \ldots 2(n)+3$

$$= 2(1+2+ \ldots n)+3n$$
$$= 2n(n+1)/2+3n \qquad \text{(Summing the A.P.)}$$
$$= n^2+4n$$

The arithmetical progression

$$a+a+d+a+2d+ \ldots a+(n-1)d$$

can be written

$$\sum_0^{n-1} (a+rd) \quad \text{or} \quad \sum_1^n [a+(r-1)d],$$

and the geometrical progression

$$a+ar+ar^2+ar^3+ \ldots +ar^{n-1}$$

can be written

$$\sum_{k=1}^{k=n} ar^{k-1} \quad \text{or} \quad \sum_{k=0}^{k=n-1} ar^k$$

Exercise 1.3

Find the value of the following:

(a) $\sum_1^4 (2r^2+1)$ (b) $\sum_2^5 \frac{r+1}{r-1}$ (c) $\sum_1^4 \frac{r^2-1}{r^2+1}$ (d) $\sum_0^3 \pi r^2$

(e) $\sum_1^3 \frac{4n^2-3}{n}$

Some Important Identities

(i) $\Sigma bx_r = bx_1 + bx_2 + bx_3 + \ldots + bx_n = b\Sigma x_r$

(ii) $\Sigma(a+bx_r) = (a+bx_1)+(a+bx_2)+ \ldots +(a+bx_n)$
$= na + b\Sigma x_r$

If there is no ambiguity, we can leave out the suffix and write the identities as

$$\text{(i) } \Sigma bx = b\Sigma x \qquad \text{(ii) } \Sigma(a+bx) = an + b\Sigma x$$

Similarly

$$\Sigma(ax+by^2-cz) = a\Sigma x + b\Sigma y^2 - c\Sigma z$$

EXAMPLE (i)

Expand $\sum_1^4 x_r^2$

SOLUTION

$$\sum_1^4 x_r^2 = x_1^2 + x_2^2 + x_3^2 + x_4^2$$

EXAMPLE (ii)

Expand $\sum_1^5 f_r x_r y_r$

SOLUTION

$$\sum_1^5 f_r x_r y_r = f_1 x_1 y_1 + f_2 x_2 y_2 + f_3 x_3 y_3 + f_4 x_4 y_4 + f_5 x_5 y_5$$

EXAMPLE (iii)

Two variables x and y have the following values:

$$x_1 = 3, \quad x_2 = -1, \quad x_3 = 4, \quad x_4 = -2$$
$$y_1 = 2, \quad y_2 = -3, \quad y_3 = 6, \quad y_4 = -4$$

Calculate:

(a) Σx (b) Σy (c) Σxy (d) Σx^2 (e) Σy^2 (f) $\Sigma x \times \Sigma y$
(g) $\Sigma 3xy^2$ (h) $\Sigma(3x+4y)$ (i) $\Sigma(x+y-3xy)$ (j) $\Sigma(x+a)^2$
(k) $\Sigma(x-a)(x+a)$

SOLUTION

(a) $\Sigma x = 3-1+4-2 = 4$
(b) $\Sigma y = 2-3+6-4 = 1$
(c) $\Sigma xy = (3)(2)+(-1)(-3)+(4)(6)+(-2)(-4)$
$\qquad = 6+3+24+8 = 41$
(d) $\Sigma x^2 = (3)^2+(-1)^2+(4)^2+(-2)^2 = 30$
(e) $\Sigma y^2 = (2)^2+(-3)^2+(6)^2+(-4)^2 = 65$
(f) $\Sigma x \times \Sigma y = 4 \times 1 = 4$ (see (a) and (b))
(g) $\Sigma(3xy^2) = 3\Sigma(xy^2) = 3[3(2)^2+(-1)(-3)^2+4(6)^2+(-2)(-4)^2]$
$\qquad = 345$
(h) $\Sigma(3x+4y) = \Sigma(3x)+\Sigma(4y) = 3\Sigma x+4\Sigma y = 16$
(i) $\Sigma(x+y-3xy) = \Sigma x+\Sigma y-3\Sigma(xy) = 4+1-3\times41 = -118$
(j) $\Sigma(x+a)^2 = \Sigma(x^2+2ax+a^2) = \Sigma x^2+2a\Sigma x+\Sigma a^2$
$\qquad = 30+2a4+4a^2 = 30+8a+4a^2$
(k) $\Sigma(x+a)(x-a) = \Sigma(x^2-a^2) = \Sigma x^2-4a^2 = 30-4a^2$

EXAMPLE (iv)

Use the Σ notation to express the following:

(a) $x_1y_1+x_2y_2+ \ldots +x_ny_n$
(b) $(x_1+y_1)^2+(x_2+y_2)^2+ \ldots +(x_n+y_n)^2$

SOLUTION

(a) $\displaystyle\sum_1^n x_ry_r$ (b) $\displaystyle\sum_1^n (x_r+y_r)^2$

Exercise 1.4

1   Express the following fully:

(a) $\sum_{1}^{4} x_r^2$ (b) $\sum_{1}^{3} x_r^3 y_r^2$ (c) $\sum_{1}^{4} (x_r+3)$ (d) $\sum_{1}^{3} (3x_r+1)$

(e) $\sum_{1}^{5} (3x_r-4y_r^2)$ (f) $\sum_{1}^{3} (x_r-1)^2$ (g) $\sum_{1}^{5} (3x_r+2y_r)^2$

2   Use the Σ notation to express the following:

(a) $x_1^2+x_2^2+ \ldots +x_7^2$
(b) $2x_1+y_1+2x_2+y_2+ \ldots +2x_{10}+y_{10}$
(c) $f_1x_1+f_2x_2+ \ldots +f_nx_n$
(d) $3f_1x_1^2+3f_2x_2^2+ \ldots +3f_nx_n^2$
(e) $f_1x_1^2y_1+f_2x_2^2y_2+ \ldots +f_nx_n^2y_n$

3   If $\sum_{1}^{10} x = 10$; $\sum_{1}^{10} y = 4$; $\sum_{1}^{10} x^2 = 150$; $\sum_{1}^{10} y^2 = 200$; find the value of:

(a) $\sum_{1}^{10} (3x+y)$ (b) $\sum_{1}^{10} (2x+2y+4x^2-2y^2)$

(c) $\sum_{1}^{10} (2y+4)$ (d) $3\sum_{1}^{10} x(x+3)$ (e) $\sum_{1}^{10} (2x+3)^2$

4   Prove $\sum_{1}^{7} (3x_r+2y_r-1)^2 = 6307$ if $\Sigma x_r = 28$; $\Sigma y_r = 56$; $\Sigma x_r^2 = 140$;

$\Sigma y_r^2 = 560$; $\Sigma x_r y_r = 266$

5   Prove $\sum_{1}^{n} (x+3)(y-1) = \sum_{1}^{n} xy - \sum_{1}^{n} x + 3\sum_{1}^{n} y - 3n$

6   Find the value of the following:

(a) $\sum_{1}^{n} (3+x_r)$ (b) $\sum_{1}^{n} (a+bx_r)$ (c) $\sum_{1}^{n} (a+b)x_r$ (d) $\sum_{1}^{n} r$.

(e) $\sum_{1}^{n} a^r$ (f) $\sum_{1}^{n} (2+r)$ (g) $\frac{1}{n}\sum_{1}^{n} (3+r)$

An Important Example

If

$$s_{xy} = \frac{\Sigma xy}{n} - \frac{\Sigma x}{n} \times \frac{\Sigma y}{n}$$

$$s_x^2 = \frac{\Sigma x^2}{n} - \left(\frac{\Sigma x}{n}\right)^2$$

$$s_y^2 = \frac{\Sigma y^2}{n} - \left(\frac{\Sigma y}{n}\right)^2$$

and

$$x_r = a + bX_r$$
$$y_r = c + dY_r$$

prove

$$s_{xy} = bd s_{XY}$$
$$s_x^2 = b^2 s_X^2$$
$$s_y^2 = d^2 s_Y^2$$

and therefore $s_{xy}/s_x s_y = s_{XY}/s_X s_Y$

Also if $\bar{x}$, $\bar{y}$, $\bar{X}$, $\bar{Y}$ are the arithmetic means of x, y, X, Y respectively, then

$$\bar{x} = a + b\bar{X}$$
$$\bar{y} = c + d\bar{Y}$$

The student is advised to try the proof of this example himself. If he finds difficulty with the proof he will find the solution in Mathematical Note 1.1. These results will be used later.

Mathematical Note 1.1.

$$\frac{\Sigma xy}{n} = \frac{1}{n}\Sigma(a+bX)(c+dY)$$

$$= \frac{1}{n}\Sigma\,(ac + adY + bcX + bdXY)$$

$$= ac + ad\frac{\Sigma Y}{n} + bc\frac{\Sigma X}{n} + bd\frac{\Sigma XY}{n}$$

$$\frac{\Sigma x}{n} = \frac{\Sigma(a+bX)}{n} = a + b\frac{\Sigma X}{n}$$

$$\frac{\Sigma y}{n} = \frac{\Sigma(c+dY)}{n} = c + d\frac{\Sigma Y}{n}$$

Therefore
$$s_{xy} = ac + ad\frac{\Sigma Y}{n} + bc\frac{\Sigma X}{n} + bd\frac{\Sigma XY}{n}$$

$$-\left(a + b\frac{\Sigma X}{n}\right)\left(c + d\frac{\Sigma Y}{n}\right)$$

$$= ac + ad\frac{\Sigma Y}{n} + bc\frac{\Sigma X}{n} + bd\frac{\Sigma XY}{n}$$

$$- ac - ad\frac{\Sigma Y}{n} - bc\frac{\Sigma X}{n} - bd\left(\frac{\Sigma X}{n} \times \frac{\Sigma Y}{n}\right)$$

$$= bd\left[\frac{\Sigma XY}{n} - \left(\frac{\Sigma X}{n} \times \frac{\Sigma Y}{n}\right)\right]$$

$$= bds_{XY}$$

$$\frac{\Sigma x^2}{n} = \frac{\Sigma(a + bX)^2}{n} = a^2 + 2ab\frac{\Sigma X}{n} + b^2\frac{\Sigma X^2}{n}$$

Therefore
$$s_x^2 = a^2 + 2ab\frac{\Sigma X}{n} + b^2\frac{\Sigma X^2}{n} - \left(a + b\frac{\Sigma X}{n}\right)^2$$

$$= b^2\left[\frac{\Sigma X^2}{n} - \left(\frac{\Sigma X}{n}\right)^2\right]$$

$$= b^2 s_X^2$$

Similarly
$$s_y^2 = d^2 s_Y^2$$

Therefore
$$s_{xy}/s_x s_y = bds_{XY}/bs_X ds_Y = s_{XY}/s_X s_Y$$

Also
$$s_{xy}/s_x^2 = bds_{XY}/b^2 s_X^2 = ds_{XY}/bs_X^2$$

and similarly
$$s_{xy}/s_y^2 = bs_{XY}/ds_Y^2$$

If $b = d$
$$s_{xy}/s_x^2 = s_{XY}/s_X^2$$

and
$$s_{xy}/s_y^2 = s_{XY}/s_Y^2$$

$$\bar{x} = \frac{\Sigma x}{n} = \frac{\Sigma(a + bX)}{n} = \frac{na + b\Sigma X}{n}$$

$$= a + b\frac{\Sigma X}{n} = a + b\bar{X}$$

Similarly
$$\bar{y} = c + d\bar{Y}$$

CHAPTER 2

Frequency Distributions

Classification

When raw data are collected they are distributed into classes, or categories, or intervals and the number of times the items appear in a class-interval is known as the class-frequency. For easy comparison it is advantageous if the class-intervals are equal, but if some of the class-frequencies are relatively small, it is more convenient to classify the data in larger intervals. A tabular arrangement of the class-intervals and the class-frequencies is known as a frequency distribution.

Class-Intervals

It is important to be able to decide the true limits of a class-interval and, to allow for this, the degree of accuracy used in measurements should be given. Thus if the heights of a number of boys are given in three cm intervals as say 152 and less than 155, 155 and less than 158, and so on; and the measurements are made correct to the nearest 0·2 of a cm, the limits of the class-intervals would be 151·9 to 154·9; 154·9 to 157·9; and so on; and if measured correct to the nearest 0·1 of a cm they would be 151·95 to 154·95; 154·95 to 157·95; and so on. The smaller measurement is known as the lower class-boundary and the larger measurement as the upper class-boundary. If an interval is given as, say, 167 and over, and there is no indication of the upper class-limit, it is known as an open interval.

In our example if the heights are measured to the nearest cm, the class-interval 161 cm includes all measurements from 160·5000 . . . to 161·5 cm, or briefly 160·5 to 161·5 cm. If the class-boundaries are used to indicate the class-intervals, there are times when ambiguity enters into the classification. Thus if the class-boundaries are given as 160·5 cm to 161·5 cm; 161·5 cm to 162·5 cm; and so on, instructions should be given to indicate in which class-interval an exact measurement such as 161·5 cm is to be placed. It is

9

preferable to name the class-intervals as 160·5 and less than 161·5; 161·5 and less than 162·5; and so on. The mid-point of the class-interval is taken as the representative point of the class-interval, and is known as the class mark or class mid-point.

Procedure for Tabulating a Frequency Distribution

  (i) Fix the magnitude of the class-interval.
 (ii) If possible allow for a maximum of about 20 to 25 class-intervals.
(iii) Classify the items according to (i) and (ii).
(iv) Draw up a table showing the frequency of each class-interval.

Histograms and Frequency Polygons

A histogram is formed by marking along the horizontal axis lengths proportionate to the class-interval, and on each length a rectangle is constructed whose *area* is proportionate to the class-frequency appropriate to the class-interval. If the class-intervals are equal, the height of the rectangle will be proportionate to the class-frequency. If the class-intervals are unequal, the areas of the rectangles above each class-interval must still be proportionate to the class-frequencies. The height of each rectangle will, therefore, be proportionate to the class-frequency divided by the class-interval.

A frequency polygon is obtained by erecting ordinates at the centre of each class-interval so that the product of the ordinate and the class-interval is proportionate to the class-frequency. The procedure is the same as for a histogram except that the height of each rectangle is represented by an ordinate at the mid-point of the class-interval. The tops of the ordinates are then joined by straight lines. If appropriate, the last points at each end are joined to the base at the centre of the next class-interval.

Relative Frequency Distributions

The relative frequency of a class is the frequency of that class divided by the sum of the frequencies of all the classes. The sum of all the relative frequencies is therefore one. If the frequencies are represented as a percentage of the total frequency the sum of the relative frequencies is 100%.

If these frequencies are tabulated, the table is called a relative frequency distribution or percentage frequency distribution. The corresponding diagrammatic representations are known as relative frequency or percentage frequency histograms or polygons.

Frequency Curves

If the number of class-intervals is increased indefinitely, and each class-interval remains finite, the outline of the histogram, or frequency polygon, approximates to a curve. Its fundamental property is that the area between any two ordinates is proportionate to the number of observations falling between the corresponding values of the variable.

Types of Frequency Curves

The following are some of the more common types of frequency curves.

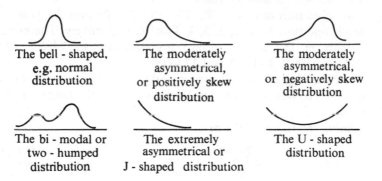

The bell - shaped, e.g. normal distribution	The moderately asymmetrical, or positively skew distribution	The moderately asymmetrical, or negatively skew distribution
The bi - modal or two - humped distribution	The extremely asymmetrical or J - shaped distribution	The U - shaped distribution

DIAGRAM 2.1

EXAMPLE

Male Deaths in the United Kingdom (1960) Classified by Age at Death

Age	0–9,	10–19,	20–34,	35–54,	55–74,	75 and over
Frequency	14 971	2 455	5 776	35 034	144 690	106 149

(A)

(i) Calculate the class boundaries
(ii) Calculate the mid-points of the class-intervals
(iii) Calculate the heights of the rectangles for the histogram
(iv) Draw the histogram and frequency polygon

(v) Calculate the relative frequencies
(vi) Name the type of distribution

SOLUTION

(i) No indication is given as to whether the age is given to the nearest year or to the nearest completed year. It is usual to give ages in completed years, so the class-intervals are better stated as 0 and under 10; 10 and under 20; 20 and under 35; 35 and under 55; 55 and under 75; 75 and over.

(ii) The sizes of the first five class-intervals are now 10, 10, 15, 20 and 20 years, respectively. The last interval is open-ended and we must decide a suitable size. If we take it as 30 years we shall be very near the truth. The mid-points of the class-intervals are therefore 5, 15, 27·5, 45, 65 and 90 years.

(iii) The class-intervals are 10, 10, 15, 20, 20 and 30 years, and are proportionate to 2, 2, 3, 4, 4 and 6, respectively. Therefore the heights of the rectangles are proportionate to 7485·5, 1227·5, 1925·3, 8758·5, 36 172·5 and 17 691·5.

To allow the histogram to fit conveniently into the page a suitable scale for the class-intervals is 0·8 cm for 10 years and for the heights 1 cm for 5000 units.

Tabulating the results we have:

Class-interval (years)	Length on graph (cm)	Class-frequency	Proportionate length of rect.	Length on graph (cm)
0 and under 10	0·8	14 971	7 485·5	1·49
10 and under 20	0·8	2 455	1 227·5	0·25
20 and under 35	1·2	5 776	1 925·3	0·39
35 and under 55	1·6	35 034	8 758·5	1·75
55 and under 75	1·6	144 690	36 172·5	7·23
75 and over	2·4	106 149	17 691·5	3·54

Diagram 2.2. shows how difficult it is to illustrate correctly a frequency distribution. The deaths are not evenly distributed in a class-interval as is suggested by the histogram, and in the frequency polygon the area above any class-interval is not exactly proportionate to the frequency in that class-interval. Since the number of deaths decreases with age in the range 0 to 15, it seems more appropriate to continue the broken line of the frequency polygon upwards from the top of the ordinate at 5 years to the ordinate at 0 years as illustrated. Drawing it downwards to −5 years, the mid-point of the preceding class-interval, would have no meaning.

(iv) Male Deaths in the United Kingdom 1960

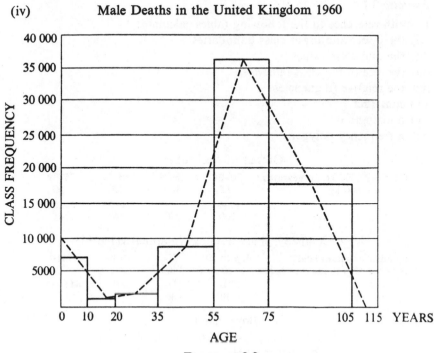

DIAGRAM 2.2

(v) The total number of deaths is 309 075.

Class-interval	Relative frequency	Relative percentage frequency
0 and under 10 years	$\frac{14\,971}{309\,075} = 0\cdot0484$	4·84
10 and under 20 years	$\frac{2455}{309\,075} = 0\cdot0079$	0·79
20 and under 35 years	$\frac{5576}{309\,075} = 0\cdot0180$	1·80
35 and under 55 years	$\frac{35\,034}{309\,075} = 0\cdot1134$	11·34
55 and under 75 years	$\frac{144\,690}{309\,075} = 0\cdot4681$	46·81
75 and over	$\frac{106\,149}{309\,075} = 0\cdot3434$	34·34

(vi) The frequency distribution is bi-modal.

Exercise 2.1

1   With reference to the following tables calculate:
 (i) the upper and lower class boundaries
 (ii) the mid-class values
(iii) the size of the class-intervals
(iv) the relative frequencies
and construct
 (v) a histogram
(vi) a frequency polygon

(a)   Output of steel tubes

Price per 1000 kg (to nearest £)	20–24	25–29	30–34	35–39
Output (1000 kg)	65	140	270	350
	40–44	45–49	50–54	55–59
	260	230	140	45

(b)   XYZ Company, share-holdings (nominal value £1)

Number of shares held	Up to 50	51–	101–	151–
Number of persons with shares	80	140	112	77
	201–	251–	401–	451–600
	50	40	9	12

(c)   Boys' masses

Mass in kg (to nearest 0·1 kg)	Number of Boys
29·5 and under 34·0	12
34·0 and under 38·5	443
38·5 and under 43·0	669
43·0 and under 47·5	820
47·5 and under 52·0	635
52·0 and under 56·5	483
56·5 and under 61·0	207
61·0 and under 65·5	72
65·5 and under 70·0	43
70·0 and under 74·5	12
74·5 and over	8

(d)   Hourly wage-rates, 1969

Hourly wage-rates (to nearest £0·1)	0·3–	0·4–	0·5–	0·6–	0·7–	0·8–	0·9–
Frequency	141	311	511	585	454	358	214
	1·0–	1·1–	1·2 to 1·5				
	131	65	59				

(e)   Masses of adult males
                        (To nearest 0·1 kg)

Mass (kg)	40–	55–	60–	65–	70–	75–	85–120
Frequency	161	1032	1240	1075	881	796	367

CHAPTER 3

Averages

Averages

When it is necessary to make comparisons between groups of numbers, it is convenient to have one figure that is representative of each group. This figure is called the average of the group. If the numbers of the group are arranged in order of magnitude, the averages tend to lie centrally in the group, so averages are sometimes called measures of central tendency. There are many averages in use, and the choice of average depends upon which best represents the property under discussion.

The Arithmetic Mean

This is the average of everyday use, and is defined as the sum of the magnitudes of the items divided by the number of items. The arithmetic mean of a set of n numbers $x_1, x_2, \ldots x_n$ is generally denoted by A.M., m, or $\bar{x}$ (read x bar) and is defined as

$$\bar{x} = \frac{x_1 + x_2 + \ldots + x_n}{n} = \frac{\Sigma x}{n}$$

if the numbers $x_1, x_2, \ldots x_k$ occur $f_1, f_2, \ldots f_k$ times, respectively

$$\bar{x} = \frac{f_1 x_1 + f_2 x_2 + \ldots + f_k x_k}{f_1 + f_2 + \ldots + f_k}$$

$$= \frac{\Sigma fx}{\Sigma f} = \frac{\Sigma fx}{n} \tag{3.1}$$

since $\Sigma f = n$.

The calculation of $\bar{x}$ can be simplified by using a working mean and the

15

deviations from the working mean. Thus if a is a working, or assumed mean, and d_r is the deviation of x_r from a, then

$$d_r = x_r - a$$

or

$$x_r = a + d_r$$

Therefore

$$\bar{x} = \frac{\Sigma fx}{n} = \frac{\Sigma f(a+d)}{n}$$

$$= \frac{\Sigma fa}{n} + \frac{\Sigma fd}{n} = \frac{a\Sigma f}{n} + \frac{\Sigma fd}{n}$$

$$= a + \frac{\Sigma fd}{n} \qquad (\Sigma f = n) \tag{3.2}$$

or the arithmetic mean is the working mean plus the average of the deviations from the working mean. If the items of a frequency distribution are classified in intervals, we make the assumption that every item in an interval has the mid-value of the interval. The error introduced is negligible if the frequencies are reasonably large.

Formula (3.1) can be used to calculate $\bar{x}$ for a frequency distribution classified in intervals if d is taken as the deviation of the mid-point of the interval from the assumed or working mean. The class-intervals will probably be of the same size, or have a common factor in the deviations. Let this common factor be c and let $d = cu$. Then

$$\bar{x} = a + \frac{\Sigma fd}{n} = a + \frac{\Sigma fcu}{n} = a + c\frac{\Sigma fu}{n} \tag{3.3}$$

This arrangement greatly simplifies the calculation of $\bar{x}$.

EXAMPLE

Use the following frequency table to find the mean height of a species of plants.

Height (to nearest cm)	60–62	63–65	66–68	69–71	72 and over
Frequency	6	18	44	25	10

Table 3.1

SOLUTION

We can proceed in two ways.

(i) Use Formula (3.1)

The intervals are really 59·5 to 62·5; 62·5 to 65·5; and so on.

Height (cm)	Mid-value of interval (x)	Frequency (f)	fx
60–62	61	6	366
63–65	64	18	1152
66–68	67	44	2948
69–71	70	25	1750
72–76 (say)	74	10	740
		$\Sigma f = 103$	$\Sigma fx = 6956$

Therefore, using Formula (3.1)

$$\bar{x} = \frac{\Sigma fx}{\Sigma f} = \frac{6956}{103} = 67 \cdot 53 \text{ cm}$$

(ii) Use Formula (3.3).

Height (cm)	Mid-value of interval (x)	Frequency (f)	Deviation from working mean (a) in units of $3(c)$ u	fu
60–62	61	6	-2	-12
63–65	64	18	-1	-18
66–68	$67 = a$	44	0	0
69–71	72	25	1	25
72–76	74	10	$2\frac{1}{3}$	$23\frac{1}{3}$
		$\Sigma f = 103$		$\Sigma fu = 18\frac{1}{3}$

Therefore, using Formula (3.3)

$$\bar{x} = a + c\frac{\Sigma fu}{\Sigma f}$$

$$= 67 + \frac{3 \times 18\frac{1}{3}}{103}$$

$$= 67 \cdot 53 \text{ cm}$$

An Important Theorem

The sum of the deviations of the items from the arithmetic mean is zero.

$$\bar{x} = \frac{\Sigma x}{n}$$

or

$$\Sigma x = n\bar{x}$$

Therefore

$$\Sigma(x - \bar{x}) = \Sigma x - n\bar{x}$$
$$= n\bar{x} - n\bar{x}$$
$$= 0.$$

The Median. Discrete Distribution

Arrange the items of a distribution in order of magnitude starting with either the largest or the smallest, then

(i) if the number of items is odd, the median is the value of the middle item,

(ii) if the number of items is even, the median is the arithmetic mean of the two middle items.

EXAMPLE

Calculate the median of

(i) 2   5   6   8   13   15   19   22   38
(ii) 3   4   8   9   15   17   18   28   40   47

SOLUTION

The items are already in ascending order.

(i) There are 9 items. Therefore the median is the fifth item, i.e. 13.
(ii) There are 10 items. Therefore the median is the average of the fifth and sixth items, i.e. $\frac{1}{2}(15+17) = 16$.

If there are n items in the distribution the median is the magnitude of the $\frac{1}{2}(n+1)$th item. This gives us the same median value whether the items are arranged in ascending or descending order of magnitude.

Continuous Distributions

The median is easily calculated for a small discrete distribution, but in the case of a distribution where the variable is continuous and classified, it is more accurate to find the median graphically. To do this, find the cumulative frequency formed by adding each class-frequency to the sum of the previous class-frequencies. Plot the cumulative frequency against the correct value of the variable and draw a smooth curve through the points. The value of the middle item can be read from the graph. In this case the median is the value of the $\frac{1}{2}n$th item because we get the same value if we cumulate the frequency by starting either with the largest or smallest item.

EXAMPLE

Find graphically the median of the distribution of Table 3.1. (p. 16)

SOLUTION

Measurements are made to the nearest cm and therefore the upper limit of the first class-interval is 62·5, the second class-interval 65·5, and so on. Therefore 6 is the total frequency to 62·5 cm, and 6+18 = 24 the total frequency to 65·5 cm, and so on.

Arranging in the form of a table we have,

Upper class limit	Frequency	Cumulative frequency
62·5	6	6
65·5	18	6+18 = 24
68·5	44	24+44 = 68
71·5	25	68+25 = 93
76·5	10	93+10 = 103

Plotting the cumulative frequency against the upper class limit, and drawing a smooth curve through the points we have the graph,

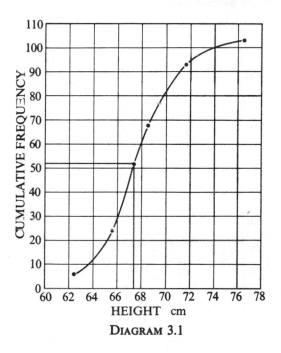

HEIGHT cm

DIAGRAM 3.1

The value of the median is the height corresponding to the 51·5th item, that is 67·4 cm.

We draw a smooth curve through the points because we assume the distribution is a random sample of plants, and, if we were considering the whole population of plants, the distribution would be continuous, and our graph would approximate to a smooth curve. The curve is called an Ogive (pronounced ōjīve). If we approximate to the curve by joining the points by straight lines, we can obtain an approximate value of the median by simple proportion.

Let Diagram 3.2 represent a cumulative frequency graph with the points joined by straight lines. Let

L = the value of the variable at the lower class boundary of the median class

n = total frequency

$(\Sigma f)_L$ = the cumulative frequency up to the lower median class boundary

f = the frequency of the median class

$\frac{1}{2}n$ = cumulative frequency up to the median

c = size of the median class

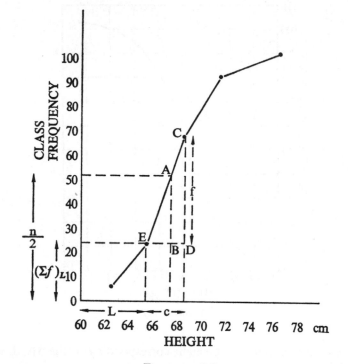

Diagram 3.2

By similar triangles

$$\frac{AB}{CD} = \frac{EB}{ED}$$

or

$$EB = \frac{AB \times ED}{CD}$$

Now

$$AB = \tfrac{1}{2}n - (\Sigma f)_L, \qquad ED = c, \qquad CD = f$$

$$EB = \frac{\tfrac{1}{2}n - (\Sigma f)_L}{f} \times c$$

and the median $= L + EB$

In our example $L = 65 \cdot 5$, $n = 103$, $(\Sigma f)_L = 24$, $f = 44$, and $c = 3$.

$$EB = \frac{\tfrac{1}{2}(103) - 24}{44} \times 3$$

$$= 1 \cdot 88$$

The median $= L + EB = 65 \cdot 5 + 1 \cdot 88 = 67 \cdot 4$.

The Mode

The mode of a distribution is the most frequent, or most 'popular' item. It may not exist and even if it does it may not be unique.

If the variable is discrete the mode can generally be easily distinguishable, but in the case of a continuous variable with class-intervals extending over a range it is more difficult. The class-interval with the largest frequency is called the modal class, and if the histogram of the distribution is symmetrical the mid-value of the modal class is taken as the mode. If the distribution is slightly asymmetrical it is better to assume the mode is that value of the variable that divides the modal class in the ratio of the difference of the frequencies of the modal class and the next lower class to the difference of the frequencies of the modal class and the next higher class.

Thus if $\quad L = $ lower class boundary of the modal class

$\Delta_1 = $ difference between the frequency of the modal class and the frequency of the next lower class

$\Delta_2 = $ difference between the frequency of the modal class and the frequency of the next upper class

$c = $ size of the modal class

then the $\quad$ Mode $= L + c\Delta_1/(\Delta_1 + \Delta_2)$

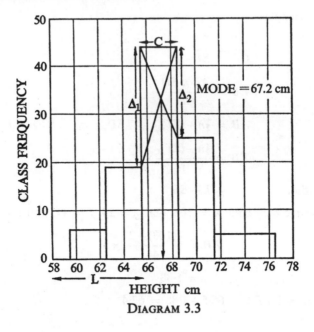

DIAGRAM 3.3

In the case of the previous example $L = 65{\cdot}5$, $\Delta_1 = 44 - 18 = 26$, $\Delta_2 = 44 - 25 = 19$, $c = 3$.

$$\text{Mode} = 65{\cdot}5 + \frac{26}{26+19} \times 3$$

$$= 67{\cdot}23 \text{ cm}$$

Diagram 3.3 illustrates how the properties of similar triangles can be used to determine the mode graphically.

Empirical Relation between the Mean, Mode and Median

For unimodal frequency curves that are only slightly asymmetrical the following empirical relation is approximately true,

$$\text{Mode} = \text{Mean} - 3(\text{Mean} - \text{Median})$$

The relative positions of the mean, median and mode in a positively and a negatively skewed distribution are shown in Diagrams 3.4 and 3.5

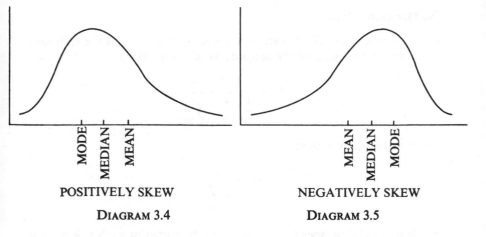

POSITIVELY SKEW

DIAGRAM 3.4

NEGATIVELY SKEW

DIAGRAM 3.5

The Geometric Mean

The geometric mean of a set of numbers $x_1, x_2, \ldots x_n$ is the nth root of the product of the numbers. Thus if G stands for the geometric mean, then

$$G = \sqrt[n]{(x_1 \times x_2 \times \ldots \times x_n)}$$

Taking logarithms, we have

$$\log G = \frac{1}{n} (\log x_1 + \log x_2 + \ldots + \log x_n)$$

Therefore the logarithm of the geometric mean is the arithmetic mean of the logarithms of the numbers.

If f_r is the frequency of x_r, then

$$G = \sqrt{(x^{f_1} \times x^{f_2} \ldots \times x^{f_n})}$$

and
$$\log G = \frac{1}{n} \Sigma f \log x \qquad (3.3)$$

If we have a frequency distribution in class-intervals and assume the mid-point of each class-interval is the average value of the variables in that class-interval, we can approximate very closely to the geometric mean of the distribution by means of formula (3.3), where x is the mid-value of a class-interval and f is the frequency of the interval. The advantage of this average over the arithmetic mean is that it reduces the effect of a single large item on the average.

The Harmonic Mean

The harmonic mean, H, of a set of numbers $x_1, x_2, \ldots x_n$ is the reciprocal of the arithmetic mean of the reciprocals of the numbers.

$$H = \frac{1}{\frac{1}{n}\sum\frac{1}{x}} = \frac{n}{\sum\frac{1}{x}}$$

If x_r has frequency f_r then

$$H = \frac{n}{\sum\frac{f}{x}}$$

Problems involving 'speeds' or 'rates' can be stated in such a way as to make the harmonic mean the appropriate average to use.

Quantiles

When a distribution is arranged in order of magnitude of the items, the median is the value of the middle term. Other measures that depend upon their position in the distribution are quartiles, deciles and percentiles. Quartiles divide the frequency of the distribution into four equal parts. The values of the variable corresponding to these divisions, generally denoted by Q_1, Q_2, and Q_3 are called the first, second, and third quartiles respectively. The second quartile is the median. The values of the variable that divide the frequency of the distribution into ten equal parts are known as deciles, and are denoted by $D_1, D_2, \ldots D_9$, while the values dividing the frequency of the distribution into one hundred equal parts are known as percentiles, and are generally denoted by $P_1, P_2, \ldots P_{99}$. The fifth decile and the fiftieth percentile is the median. Quartiles, deciles and percentiles are called, collectively, quantiles. They can all be found from the ogive or by calculation.

EXAMPLE 1

Calculate the averages of the following set of fifteen numbers.

$$1 \quad 3 \quad 13 \quad 6 \quad 8 \quad 14 \quad 8 \quad 9 \quad 8 \quad 12 \quad 21 \quad 18 \quad 9 \quad 10 \quad 8$$

SOLUTION

Arranging in order of magnitude we have,

1 3 6 8 8 8 8 9 9 10 12 13 14 18 21 Total = 148

The arithmetic mean $= \dfrac{\Sigma x}{n} = \dfrac{148}{15} = 9{\cdot}867$

The median is the $\left(\dfrac{15+1}{2}\right)$th item = 8th item = 9

The mode = 8

The geometric mean $= \sqrt[15]{(1 \times 3 \times 6 \times 8^4 \times 9^2 \times 10 \times 12 \times 13 \times 14 \times 18 \times 21)}$
$= 8{\cdot}2$

The harmonic mean

$$\Sigma \frac{1}{x} = 1 + \tfrac{1}{3} + \tfrac{1}{6} + \tfrac{4}{8} + \tfrac{2}{9} + \tfrac{1}{10} + \tfrac{1}{12} + \tfrac{1}{13} + \tfrac{1}{14} + \tfrac{1}{18} + \tfrac{1}{21}$$

$$= 2{\cdot}657$$

$$H = \frac{15}{2\ 657} = 5{\cdot}65$$

The lower quartile is the $\dfrac{(n+1)}{4}$th item = 4th item = 8

EXAMPLE 2

Calculate the various averages for the following distribution.

Masses of 100 randomly chosen potatoes

Mass (nearest g)	118–126	127–135	136–144	145–153
Frequency	8	13	21	27
	154–162	163–171	172–180	
	13	11	7	

SOLUTION

Arranging the required information in tabular form we have:

Col. 1	Col. 2	Col. 3	Col. 4	Col. 5	Col. 6	Col. 7
			Deviation			
Mass (g)	Mid-class x	Frequency f	from W.M. (a) in units of 9 (c)	fu	Upper class boundaries	Cumulative frequency
118–126	122	8	−3	−24	126·5	8
127–135	131	13	−2	−26	135·5	21
136–144	140	21	−1	−21	144·5	42
145–153	149 = a	27	0	0	153·5	69
154–162	158	13	1	13	162·5	82
163–171	167	11	2	22	171·5	93
172–180	176	7	3	21	180·5	100
		$\Sigma f = 100$		$\Sigma fu = -15$		

$$\text{Arithmetic Mean} = \bar{x} = a + c \times \frac{\Sigma fu}{n} = 149 + 9 \times \frac{-15}{100} = 147 \cdot 65 \text{ g}$$

The Median and Quantiles

The upper class boundaries are given in Col. 6 and the cumulative frequencies in Col. 7. Plotting these we have the graph in Diagram 3.6

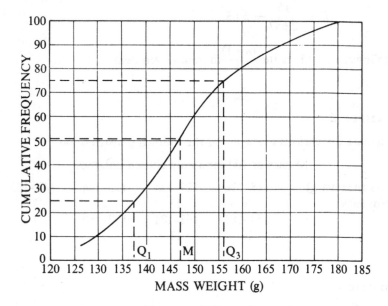

DIAGRAM 3.6

From the graph we obtain:

Median = 50th item = 147·2 Q_1 = 25th item = 137·4
Q_2 = 50th item = 147·2 Q_3 = 75th item = 156·2

D_1 = 128·6; ... D_9 = 168·2
P_{45} = 145·8 etc.

To calculate the median we have:

$$L = 144·5, \qquad n/2 = 50, \qquad (\Sigma f)_L = 42, \qquad f = 27, \qquad c = 9$$

$$\text{Median} = 144·5 + \frac{50-42}{27} \times 9 = 147·2$$

The quantiles can also be calculated by simple proportion. Thus the frequency to the first quartile is 25. Drawing a diagram for the class-interval containing the first quartile we have Diagram 3.7

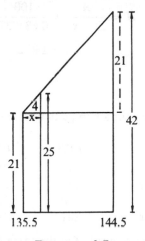

DIAGRAM 3.7

By simple proportion, we have $x/9 = 4/21$

Therefore $x = 1·71$ and $Q_1 = 135·5 + 1·7 = 137·2$

To calculate the mode, we have:

$$L = 144·5, \qquad \Delta_1 = 27-21 = 6, \qquad \Delta_2 = 27-13 = 14, \qquad c = 9$$

$$\text{Mode} = 144·5 + \frac{6}{6+14} \times 9 = 147·2$$

The geometric mean

$$= \sqrt[100]{(122^8 \times 131^{13} \times 140^{21} \times 149^{27} \times 158^{13} \times 167^{11} \times 176^7)}$$
$$= 146.93 \qquad \text{(Use 7-figure logs or a calculating machine)}$$

The harmonic mean

$$
\begin{aligned}
1/x_1 &= 0.008197 & f_1/x_1 &= 0.065576 \\
1/x_2 &= 0.007634 & f_2/x_2 &= 0.099242 \\
1/x_3 &= 0.007143 & f_3/x_3 &= 0.150003 \\
1/x_4 &= 0.006711 & f_4/x_4 &= 0.181197 \\
1/x_5 &= 0.006329 & f_5/x_5 &= 0.082277 \\
1/x_6 &= 0.005988 & f_6/x_6 &= 0.065868 \\
1/x_7 &= 0.005682 & f_7/x_7 &= 0.039774 \\
& & \Sigma(f/x) &= \overline{0.683937}
\end{aligned}
$$

$$H = \frac{n}{\Sigma(f/x)} = \frac{100}{0.683937}$$

$$= 146.2$$

EXAMPLE

The quarterly sales figures of a company over three years were, in units of £1000

	1968	1969	1970
First quarter	242	256	266
Second quarter	259	278	289
Third quarter	270	283	303
Fourth quarter	281	295	320

Plot the above figures on a graph. On the same graph plot the four-point moving averages and assuming the general trend can be represented by a straight line, draw this line.

From the graph predict the sales for the first quarter of 1971.

SOLUTION

(Note: A four-point moving average is obtained by averaging the first, second, third and fourth terms, then the second, third, fourth and fifth terms, and so on. Each average is plotted at the mid-point of the period it represents. It is a means of smoothing out unwanted variations. The span of a moving average should be equal to, or a multiple of, the number of observations in the variation to be removed.) Calculating the averages, we have

263   266·5   271·25   274·5   278   280·5   283·25   288·25   294·5

Plotting and drawing a straight line through the moving averages we have the graph below.

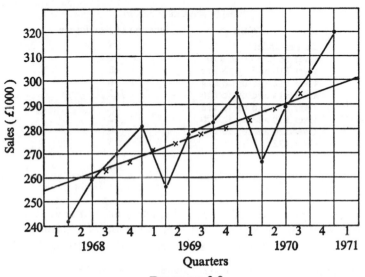

DIAGRAM 3.8

The deviations from the line of best fit for the first quarters are −16·5, −16·5 and −21 and the average deviation is −18·0. The general trend for the first quarter of 1971 is 301. Therefore, the expected sales are 301 − 18 = 283 thousand pounds.

Exercise 3.1

1   Calculate the arithmetic mean, median and mode for the following distributions.

(a) 3   4   9   13   6   9   7   12   9   100

(b)

Marks	1–10	11–20	21–30	31–40	41–50	51–60	61–70
Frequency	11	42	83	145	171	133	101

71–80	81–90	91–100
72	41	18

(c) Length of leaf (mm) 20–   25–   30–   35–   40–   45–   50–   55–   60–
    Frequency   2   7   12   21   27   23   17   5   5
Measurements are made correct to the nearest millimetre.

2   The table below shows the weekly wage distribution in a certain factory. Determine the (i) median wage, (ii) wage limits for the middle 50% of the wage earners, (iii) percentage of workers who earned more than £13 per week, (iv) percentage of workers who earned £17 or less per week.

Weekly wage distribution

Wage	Number of Employees
£8 and under	15
Over £8 and up to £10	20
Over £10 and up to £12	48
Over £12 and up to £14	76
Over £14 and up to £16	112
Over £16 and up to £18	64
Over £18 and up to £20	32
Over £20 and up to £22	10
Over £22	3
	380

(AEB)

3   A machine produces rods whose diameters are required to be within the tolerance limits 0·988 cm to 1·012 cm. A sample of 150 rods, measured to the nearest thousandth of a centimetre, gave the following distribution:

Diameter (cm)	0·976 to 0·981	0·982 to 0·987	0·988 to 0·993	0·994 to 0·999	1·000 to 1·005	1·006 to 1·011	1·012 to 1·017
Number of rods	1	5	30	71	34	7	2

(i) Construct the cumulative frequency curve for the rods.
(ii) Calculate the percentage number of rods outside the tolerance limits.

(AEB)

4   Define the arithmetic mean, the median and the mode of a distribution, and give an example in which they coincide.

In a certain factory 10 employees earn £6 per week, 35 earn £10 per week, 25 earn £20 per week and 30 earn £30 per week. Find the mean, median and the mode of this wage distribution.

Which of these will be altered if 4 employees work overtime and each increases his weekly wage by £5?

(L)

5 The following table shows the marks obtained by the same 150 candidates in two different papers in a recent examination.

Marks	0–10	11–20	21–30	31–40	41–50	51–60
No. of candidates (Paper 1)	2	6	20	28	31	20
No. of candidates (Paper 2)	3	12	22	31	33	15

	61–70	71–80	81–90	91–100
	17	11	8	7
	14	9	9	2

Estimate the mean and median marks for Paper 1 only. If you were told that the true median mark for Paper 1 was in fact 48, what conclusion could you draw about the grouping of the marks?

Comment on the relative difficulty of the two papers. (C)

6 A mass-produced circular disc should have radius 2 cm but in fact the values of the radius are uniformly distributed in the range 1·95–2·10 cm. Explain why

(a) median of the area $= \pi$ (median of the radius)^2,
(b) mean of the area $\neq \pi$ (mean of the radius)^2.

Obtain limits A_L and A_U for the area such that 20% of the discs have area $<A_L$ and 20% have area $>A_U$.

7 The following distribution of marks (out of 100) was obtained with a certain examination paper.

Mark range	10–29	30–39	40–49	50–59	60–69	70–79	80–
Frequency	2	8	14	26	28	10	4

Present these marks in a histogram.

Calculate the mean mark as well as you can, explaining the limitations of your calculation. Do you consider that the mean or the median is the better measure of average performance for these observations? Give reasons for your answer. (C)

8 Sales of footwear by a store in 52 consecutive weeks were:

37	60	67	63	69	54	68	60	62	83	66	70	68	61
74	94	87	66	69	66	98	62	78	90	47	70	68	98
40	73	93	51	70	71	56	56	58	57	47	76	59	64
46	53	54	67	80	79	77	77	49	73				

Reduce these results to a frequency distribution with intervals centred at 40, 50, 60, etc., articles. Represent the results as a histogram drawn on a scale of 1 cm to each interval.

On a separate diagram plot the frequencies up to the ends of each interval and sketch a cumulative frequency curve. From this curve read off the median and quartiles of the distribution.   (O and C)

9   Prices of six metals in pounds sterling per tonne for the years 1954, 1958 and 1959 are given in the following table together with weights proportional to the amounts spent on purchases of these metals by a manufacturer in 1954.

Metal	1954	1958	1959	Weight
Aluminium	156	184	180	5
Copper	249	198	238	15
Lead	96	73	71	8
Nickel	487	600	600	9
Tin	720	735	786	9
Zinc	78	66	82	10

Taking the year 1954 as base year, calculate the index number of price relatives

(i) for the years 1958 and 1959, using the weighted arithmetic mean,

(ii) for the year 1959, ignoring the weights and using the geometric mean.
                                                                        (O and C)

10   The lives of 50 electric lamps in hours, to the nearest hour, are given in the following table:

695	716	730	689	689	700	726	662	681	724	676	732
676	697	710	694	715	738	696	696	682	699	714	707
697	710	660	703	717	692	698	684	695	682	721	708
722	692	717	656	696	701	699	705	680	702	690	663
694	671										

Form a frequency distribution by grouping these values with a class-interval of 10 hours and draw a histogram of the distribution.

By drawing a cumulative frequency curve, or otherwise, estimate the median, mode and 8th decile of the distribution.   (O and C)

11   The average price of a certain commodity, in pounds sterling per tonne, for each quarter over the period 1957 to 1961 was as follows:

	1957	1958	1959	1960	1961
1st quarter	34·4	35·2	36·6	37·7	38·0
2nd quarter	34·1	34·9	36·6	37·3	37·6
3rd quarter	35·7	36·8	38·0	38·0	38·8
4th quarter	36·6	37·6	38·3	38·8	39·1

Plot these values on a graph and on the same graph plot a moving average with a period chosen to remove the seasonal variation.

Draw a straight line to fit the moving averages and, by considering variations of the original points from this line, obtain average values of the seasonal variations. (O and C)

12 Index of retail prices (17 Jan. 1956 = 100)

	All items	Food	Alcoholic drink	Tobacco	Housing	Other goods and services
Weights	1000	350	71	80	87	412
			Monthly averages			
1959	109·6	108·2	100·0	107·9	127·8	108·7
1960	110·7	107·4	98·2	111·9	131·7	110·3
1961	114·5	109·1	102·5	117·7	137·6	115·7

(i) Explain briefly how the index of retail prices is calculated, referring to the above table.

(ii) Verify from the above figures the monthly average retail prices index for all items for 1961.

(iii) If the average prices for 1959 were taken as a new base and weights and sub-weights were unaltered, what would be the new index of retail prices for 1961? (O and C)

13 The table gives the masses in grammes of 48 golf balls measured to the nearest centigramme:

45·61	45·08	45·00	45·26	45·42	45·38	45·38	45·50	45·45
45·05	45·38	45·47	45·29	45·29	45·42	44·79	45·20	44·76
45·34	45·66	45·34	45·29	45·48	45·31	44·92	45·46	45·41
44·98	45·69	45·37	45·21	45·55	45·44	45·68	45·02	45·53
45·04	45·02	45·37	45·28	45·34	45·47	45·15	45·52	45·38
44·95	45·57	45·14						

Divide the range of variation into equal intervals of which the first is 44·70 to 44·79 g. Calculate the frequency in each interval and draw a histogram of the masses.

Draw a cumulative frequency curve to fit the population of which these masses are a sample and from this curve read off the median mass and the values of the quartiles. (O and C)

14 The following table gives the quarterly sales at a department store in thousands of pounds over a period of four years:

	1st Qr.	2nd Qr.	3rd Qr.	4th Qr.
1959	128	142	149	170
1960	131	148	153	173
1961	134	150	158	178
1962	140	155	163	192

Plot these values on a graph and on the same graph plot a moving average to remove the seasonal variation.

Draw a straight line to fit the moving average points as closely as possible and obtain average values of the seasonal variations.   (O and C)

15   The monthly average exports of United Kingdom produce and manufactures to American and European countries are given, in units of £1M, for the years 1958 and 1962 in the following table:

	1958	1962
Canada	15·7	15·7
United States	22·9	27·5
Latin America	12·6	13·4
EFTA Countries	29·2	27·5
EEC Countries	34·9	60·0
Other Countries	17·2	27·3

Taking 1958 as the base year, calculate index numbers for the year 1962 using
(a) the simple aggregate of actual prices,
(b) the average of relative prices,
(c) the geometric mean of relative prices.   (O and C)

16   The speeds in km/h of vehicles passing along a road were measured and the results given in the following table:

Centre of interval	25	30	35	40	45	50	55	60	Total
Frequency	5	25	72	122	180	256	118	22	800

Illustrate these results by drawing (i) a histogram, (ii) a cumulative frequency curve.

From the graphs estimate the median, the semi-interquartile range and the percentage of vehicles travelling at speeds greater than 55 km/h.
   (O and C)

17   The following table gives the quarterly sales (unit £1000) of a store over a period of four years:

Year	January–March	April–June	July–September	October–December
1960	36	22	30	34
1961	42	28	42	42
1962	52	34	46	52
1963	60	46	56	62

Calculate and tabulate moving averages of the quarterly sales that will remove the seasonal variations. Plot these moving averages on a graph and draw a straight line to fit the points as nearly as possible.

Estimate the sales for the four quarters of 1964. (O and C)

18 What is meant by a time-series?

Describe the various types of movement into which changes in a time-series may be classified. Is it in general possible to distinguish absolutely between types of variation?

Illustrate your answer by using the following data giving the monthly consumption of electrical energy in suitable units, over three years:

Year	Jan.	Feb.	Mar.	Apr.	May	Jun.	Jul.	Aug.	Sep.	Oct.	Nov.	Dec.
1962	42	38	37	33	31	30	30	33	36	40	42	45
1963	45	41	40	36	34	32	34	36	39	43	45	48
1964	49	44	43	39	37	35	36	39	42	46	49	52

(AEB)

CHAPTER 4

Dispersion

Dispersion

The scatter or spread of the items of a distribution is known as dispersion, and the index giving the measure of the dispersion is known as the co-efficient of deviation. Various measures of dispersion are in use, the most common being the range, mean deviation, semi-interquartile range and the standard deviation.

The Range

The difference between the magnitude of the largest and the smallest of items of the distribution is the range. As its value depends on only two items of the distribution it is a poor measure of dispersion.

The Mean Deviation

The mean deviation is the arithmetic mean of the positive values of the deviations from their arithmetic mean. The positive value of a number is indicated by two vertical lines placed around the number. Thus the positive or absolute value of the deviation of an item x from the mean $\bar{x}$ is written $|x-\bar{x}|$, and is read as modulus $x-\bar{x}$.

EXAMPLE

Calculate the mean deviation of the numbers 3, 4, 6, 8, 10, 11.

SOLUTION

The arithmetic mean $= \dfrac{3+4+6+8+10+11}{6} = 7$

The mean deviation $= \dfrac{|3-7|+|4-7|+|6-7|+|8-7|+|10-7|+|11-7|}{6}$

$\qquad\qquad\quad = 2 \cdot 667$

Generally if x_r occurs f_r times, the mean deviation is $\dfrac{\Sigma f_r |x_r - \bar{x}|}{n}$. This formula can be used for a grouped distribution if x_r is taken as the mid-class value and f_r as the frequency of the class.

EXAMPLE

Calculate the mean deviation of Example 2 on page 25.

SOLUTION

The mean value of the masses of the potatoes was 147·65 g. Tabulating the results we have:

Mid-class	Frequency	$f\|x-\bar{x}\|$
122	8	$8\|122-147\cdot65\| = 205\cdot20$
131	13	$13\|131-147\cdot65\| = 216\cdot45$
140	21	$21\|140-147\cdot65\| = 160\cdot65$
149	27	$27\|149-147\cdot65\| = 36\cdot45$
158	13	$13\|158-147\cdot65\| = 134\cdot55$
167	11	$11\|167-147\cdot65\| = 212\cdot85$
176	7	$7\|176-147\cdot65\| = 205\cdot45$
	100	1171·60

$$\text{Mean deviation} = \frac{1171\cdot60}{100} = 11\cdot72$$

The Semi-interquartile Range or Quartile Deviation

This is half the difference between the upper and lower quartiles.

$$\text{Quartile deviation} = \tfrac{1}{2}(Q_3 - Q_1)$$

Its advantages are the ease with which it can be calculated and its clear and simple meaning.

The Root-mean-square Deviation and the Standard Deviation

The root-mean-square deviation is the square root of the average of the squares of the deviations from a working mean. Thus if f_r is the frequency of x_r and a is the working mean, then root-mean-square deviation

$$s = \sqrt{\left[\frac{\Sigma f_r(x_r - a)^2}{n}\right]}$$

If $a = \bar{x}$ the root-mean-square deviation is then called the standard deviation (σ).

$$\text{Standard deviation} = \sigma = \sqrt{\left[\frac{\Sigma f(x-\bar{x})^2}{n}\right]}$$

An Important Theorem

Let $\qquad\qquad a = \bar{x}+b$

Then $\qquad\qquad s^2 = \dfrac{\Sigma f(x-a)^2}{n}$

$$= \frac{\Sigma f(x-\bar{x}-b)^2}{n}$$

$$= \frac{\Sigma f[(x-\bar{x})-b]^2}{n}$$

$$= \frac{\Sigma f(x-\bar{x})^2 - 2b\Sigma f(x-\bar{x}) + nb^2}{n}$$

$\Sigma f(x-\bar{x})$ is the sum of the deviations of the items from their arithmetic mean, and we have proved this to be zero. (Chapter 3, page 17.)

Therefore $\qquad\qquad s^2 = \dfrac{\Sigma f(x-\bar{x})^2}{n} + b^2$

$$= \sigma^2 + b^2$$

Simplification.

$$s = \sqrt{\left[\frac{f(x-a)^2}{n}\right]}$$

If, as we did on page 16, we put $(x-a)$ equal to d and assume the d's have a common factor c, so that $x-a = d = cu$, then

$$s = \sqrt{\left(\frac{\Sigma f d^2}{n}\right)}$$

$$= \sqrt{\left(\frac{\Sigma f c^2 u^2}{n}\right)}$$

$$= c\sqrt{\left(\frac{\Sigma f u^2}{n}\right)}$$

These two relations simplify the calculation of the standard deviation. σ^2 is called the variance or variability of the distribution.

EXAMPLE

Calculate the standard deviation for the masses of potatoes in the table on page 25.

SOLUTION

Mass (g)	Mid-class x	Frequency f	Deviation from a in units of 9 (c) u	fu	fu^2
118–126	122	8	−3	−24	72
127–135	131	13	−2	−26	52
136–144	140	21	−1	−21	21
145–153	149→a	27	0	0	0
154–162	158	13	1	13	13
163–171	167	11	2	22	44
172–180	176	7	3	21	63
		$\Sigma f = 100$		$\Sigma fu = -15$	$\Sigma fu^2 = 265$

$$\bar{x} = a + c\frac{\Sigma fu}{n} = 149 + \left(9 \times \frac{-15}{100}\right) = 147 \cdot 65$$

$$s^2 = c^2\frac{\Sigma fu^2}{n} = 81 \times \frac{265}{100} = 214 \cdot 65$$

$$b = a - \bar{x} = 149 - 147 \cdot 65 = 1 \cdot 35$$

$$\sigma^2 = s^2 - b^2 = 214 \cdot 65 - 1 \cdot 8225 = 212 \cdot 8275$$

$$\sigma = 14 \cdot 57$$

Alternative Method of Calculating the Standard Deviation

$$\sigma = \sqrt{\left[\frac{\Sigma f(x - \bar{x})^2}{n}\right]}$$

$$= \sqrt{\left[\frac{\Sigma f(x^2 - 2x\bar{x} + \bar{x}^2)}{n}\right]}$$

$$= \sqrt{\left(\frac{\Sigma fx^2}{n} - \frac{2\bar{x}\Sigma fx}{n} + \frac{n\bar{x}^2}{n}\right)}$$

$$= \sqrt{\left(\frac{\Sigma fx^2}{n} - 2\bar{x}^2 + \bar{x}^2\right)} \qquad \left(\frac{\Sigma fx}{n} = \bar{x}\right)$$

$$= \sqrt{\left[\frac{\Sigma fx^2}{n} - \left(\frac{\Sigma fx}{n}\right)^2\right]}$$

EXAMPLE

Apply the method to the previous example.

Mass (g)	Mid-class x	Frequency f	fx	fx²
118–126	122	8	976	119 072
127–135	131	13	1 703	223 093
136–144	140	21	2 940	411 600
145–153	149	27	4 023	599 427
154–162	158	13	2 054	324 537
163–171	167	11	1 837	306 779
172–180	176	7	1 232	216 832

$$\Sigma f = 100 \quad \Sigma fx = 14\ 765 \quad \Sigma fx^2 = 2201\ 335$$

$$\sigma = \sqrt{\left[\frac{\Sigma fx^2}{n} - \left(\frac{\Sigma fx}{n}\right)^2\right]}$$

$$= \sqrt{[22\ 013\cdot35 - 21\ 800\cdot5225]}$$

$$= 14\cdot57$$

Properties of the Standard Deviation

(i) $s^2 = \sigma^2 + b^2$; therefore s is a minimum when $b = 0$. That is, the standard deviation is the minimum root-mean-square deviation.

(ii) If an item is picked at random from a distribution, the standard deviation of the distribution gives us a measure of the likelihood of

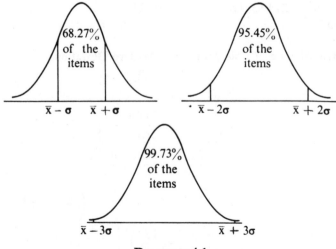

DIAGRAM 4.1

its having a value which is near the arithmetic mean of the distribution. If the distribution is the symmetrical and bell-shaped one known as the normal distribution then 68·27% of the items lie between $(\bar{x}-\sigma)$ and $(\bar{x}+\sigma)$, 95·45% between $(\bar{x}-2\sigma)$ and $(\bar{x}+2\sigma)$, and 99·73% or nearly all the items between $(\bar{x}-3\sigma)$ and $(\bar{x}+3\sigma)$ (see Diagram 4.1).

Sheppard's Correction for Variance*

If a continuous frequency distribution (or a discrete one with no gaps in the data) is classified into intervals, an error is introduced into the calculations. It is found that the mid-class values to the left of the mode tend to be smaller than the arithmetic means of the classes, and those to the right of the mode tend to be larger than the arithmetic means of the classes. Thus errors in computations involving the deviations of the class midpoints from $\bar{x}$ tend to cancel each other out and so no correction is needed when $\bar{x}$ is calculated, but errors in computations involving the squares of the deviations (as in variance) are accumulative because the signs of the squares of the deviations are all positive. The result is that the value of σ for a frequency distribution in class-intervals is usually too large. W. F. Sheppard calculated a correction for use in distributions of continuous variables. It is

Corrected Variance = Calculated Variance from Classified Data $-(c^2/12)$

where c is the class-interval size.

It should not be assumed that the general application of this correction always reduces the error in σ^2. Table 4.1 gives two distributions of marks, and Table 4.2 the results of calculating the means and standard deviations of both distributions from the ungrouped data and from grouping the data.

Distribution A is symmetrical and tapers off slightly to the left and right, while distribution B is not symmetrical and does not taper off. The results show that in Distribution A the mean is not affected by grouping and the value of σ improves as the class-interval gets smaller. Sheppard's correction only improves the value of σ when the items are grouped in intervals of 10. In Distribution B the values of the mean and σ improve as the class-interval gets smaller, while 'correcting' σ only increases the error in σ.

* For a mathematical discussion of Sheppard's Correction see *Statistical Mathematics* by A. C. Aitken, D.Sc., F.R.S. (Oliver and Boyd).

Distribution A

Marks	1	2	3	4	5	6	7	8	9	10
0	0	0	0	0	0	1	0	1	0	1
10	1	1	1	1	1	1	1	2	2	2
20	2	3	3	3	3	3	4	4	4	5
30	5	4	4	4	3	3	3	3	3	2
40	2	2	2	1	1	1	1	1	1	1
50	1	0	1	0	1	0	0	0	0	0

Distribution B

Marks	1	2	3	4	5	6	7	8	9	10
0	1	1	3	2	1	2	0	1	0	3
10	0	3	2	1	2	2	2	2	0	0
20	1	2	1	1	2	1	0	2	2	1
30	2	2	0	1	0	1	5	3	2	2
40	0	1	1	2	2	0	0	2	6	3
50	2	2	3	4	2	2	3	2	1	3

Table 4.1

Distribution	A			B		
	Mean	σ	Corrected σ	Mean	σ	Corrected σ
From ungrouped data	30·5	10·17		34	17·88	
Classified in intervals of 20	30·5	11·31	9·73	33·1	16·41	15·36
Classified in intervals of 10	30·5	10·63	10·23	33·7	17·46	17·21
Classified in intervals of 5	30·5	10·21	10·11	33·85	17·81	17·76

Table 4.2

Thus the application of Sheppard's correction needs careful consideration and should not be applied unless the distribution:

(i) is nearly symmetrical,
(ii) is continuous, or if discrete there are no gaps in the data,
(iii) tapers off left and right of the mode to zero,
(iv) the class-interval is not too large. It must be less than the standard deviation.

If these conditions are not satisfied there is a tendency to overcorrect and give a value of the variance smaller than that calculated from the original ungrouped data.

In this book Sheppard's correction will not be used unless stated.

Absolute Measures of Dispersion

The measures of dispersion discussed are all expressed in terms of units of the distributions. It is thus impossible to compare dispersions in different units. For this reason it has been suggested that 'absolute' measures of dispersion, that is measures that are pure numbers and not expressed in any unit, should be used.

The most common are:

(i) Quartile coefficient of dispersion $= \dfrac{Q_3 - Q_1}{Q_3 + Q_1}$

(ii) Coefficient of mean dispersion $= \dfrac{\text{mean absolute deviation from mean}}{\text{arithmetic mean}}$

   or, $\dfrac{\text{mean absolute deviation from median}}{\text{median}}$

   or, $\dfrac{\text{mean absolute deviation from mode}}{\text{mode}}$

(iii) Coefficient of standard dispersion $= \dfrac{\text{standard deviation}}{\text{arithmetic mean}}$

(iv) Coefficient of variation $= \dfrac{\sigma}{\bar{x}} \times 100$

The coefficient of variation is the only absolute measure of dispersion that has received favour. Obviously the coefficient becomes unreliable if $\bar{x} \to 0$, but it has a certain utility in comparing the variations of distributions of the same type.

EXAMPLE

Calculate the various coefficients of dispersion for the set of numbers,

$$5 \quad 4 \quad 3 \quad 7 \quad 12 \quad 8 \quad 6 \quad 4 \quad 3 \quad 10$$

SOLUTION

The arithmetic mean $= (5+4+3+7+12+8+6+4+3+10)/10 = 6 \cdot 2$.
Arranging the numbers in ascending order of magnitude, we have:

$$3 \quad 3 \quad 4 \quad 4 \quad 5 \quad 6 \quad 7 \quad 8 \quad 10 \quad 12$$

Median $= \frac{1}{2}(5+6) = 5 \cdot 5$
$Q_1 = 4$; $Q_3 = 8$ Quartile deviation $= \frac{1}{2}(8-4) = 2$
Range $= 12 - 3 = 9$

Mean deviation $= [|3\text{-}6\text{-}2|+|3\text{-}6\text{-}2|+|4\text{-}6\text{-}2|+|5\text{-}6\text{-}2|+|6\text{-}6\text{-}2|+|7\text{-}6\text{-}2|$
$\qquad\qquad\qquad +|8\text{-}6\text{-}2|+|10\text{-}6\text{-}2|+|12\text{-}6\text{-}2|]/(10)$
$\qquad\qquad = 2\text{-}44$

Standard deviation $= \sqrt{\left[\dfrac{\Sigma(x-\bar{x})^2}{n}\right]} = 2\text{-}90$

EXAMPLE

The table gives the frequency distribution of the volume of cells in the blood of 100 young women. Calculate the various coefficients of dispersion.

Vol. of cells (ml)	35·0–36·5	36·5–38·0	38·0–39·5	39·5–41·0
Frequency	3	7	12	26
	41·0–42·5	42·5–44·0	44·0–45·5	45·5–47·0
	28	13	9	2

SOLUTION

Note: 36·5 is the upper class limit of the first class and the lower class limit of the second class; and so on.

Upper class limit	36·5	38·0	39·5	41·0	42·5	44·0	45·5	47·0
Cum. frequency	3	10	22	48	76	89	98	100

Assuming the distribution approximates to a continuous distribution,

Median $= L+\dfrac{n/2-(\Sigma f)_L}{f}\times c = 41\text{-}0+\dfrac{50-48}{28}\times 1\text{-}5 = 41\text{-}1$

Range $= 47\text{-}0-35\text{-}0 = 12\text{-}0$

$Q_1 = 39\text{-}5+\dfrac{25-22}{26}\times 1\text{-}5 = 39\text{-}7, \qquad Q_3 = 14\text{-}0+\dfrac{75-48}{26}\times 1\text{-}5 = 42\text{-}5$

Quartile deviation $= \frac{1}{2}(42\text{-}5-39\text{-}7) = 1\text{-}4$

Mean

Mid-class		Deviation in units of 1·5	fu	fu^2	$|x-\bar{x}|$	$f|x-\bar{x}|$
35·75	3	−4	−12	48	6·69	20·07
37·25	7	−3	−21	63	5·19	36·33
38·75	12	−2	−24	48	3·69	44·28
40·25	26	−1	−26	26	2·19	56·94
41·75	28	0	0	0	0·69	19·32
43·25	13	1	13	13	0·81	10·53
44·75	9	2	18	36	2·31	20·79
46·25	2	3	6	18	3·81	7·62
	100		−46	252		215·88

Arithmetic mean $= 41{\cdot}75 - \dfrac{46 \times 1{\cdot}5}{100} = 42{\cdot}44$

Standard deviation

$$s^2 = \frac{252 \times (1{\cdot}5)^2}{100} = 5{\cdot}67, \qquad b^2 = 0{\cdot}4761, \qquad \sigma^2 = 5{\cdot}194, \qquad \sigma = 2{\cdot}28$$

$$\sigma^2 - (c^2/12) = 5{\cdot}194 - 0{\cdot}188 = 5{\cdot}006, \qquad \sigma_{\text{cor}} = 2{\cdot}24$$

Mean deviation $= \dfrac{215{\cdot}88}{100} = 2{\cdot}16$

EXAMPLE

The number of members, means and standard deviations of two distributions are:

Distribution	1	2
Number of members	200	300
Means	30	40
Standard deviations	5	6

Find the mean and standard deviation of the distribution formed by the two distributions taken together.

SOLUTION

The total number of members in the combined distribution

$$= 200 + 300$$
$$= 500$$

The sum of the values of the members in the combined distribution

$$= 200 \times 30 + 300 \times 40$$
$$= 18\ 000$$

Therefore the mean of the combined distribution

$$= \frac{18\ 000}{500}$$
$$= 36$$

Using
$$\sigma^2 = \frac{\Sigma f x^2}{n} - \left(\frac{\Sigma f x}{n}\right)^2$$

we have for Distribution 1,

$$(5)^2 = \frac{\Sigma fx^2}{200} - (30)^2$$

and $\Sigma fx^2 = 185\,000$

and for Distribution 2,

$$(6)^2 = \frac{\Sigma fx^2}{300} - (40)^2$$

and $\Sigma fx^2 = 490\,800$

Therefore for the combined distribution

$$\Sigma fx^2 = 185\,000 + 490\,800$$
$$= 675\,800$$
$$n = 200 + 300$$
$$= 500$$

and $$\sigma^2 = \frac{675\,800}{500} - (36)^2$$
$$\doteq 55 \cdot 6$$
$$\sigma = 7 \cdot 46$$

Exercise 4.1

1   Calculate the various coefficients of dispersion for the distributions of Question 1, Exercise 3.1.

2   The following figures give the numbers, in thousands, of the new school places in the ten consecutive years 1952 to 1961.

203   262   197   212   238   280   294   246   216   203

Calculate the (i) median, (ii) mean, (iii) variance of this group of numbers.

(AEB)

3   The following table gives the distribution of the marks of 1000 candidates in an examination:

Marks	1–10	11–20	21–30	31–40	41–50	51–60	61–70	71–80	81–100
No. of candidates	15	54	112	182	240	191	133	57	16

Working from an assumed mean of 45·5, calculate the mean and standard deviation of this distribution.

(AEB)

4 The table below gives the masses in grammes of 1000 articles.

Mass	Frequency	Mass	Frequency
Under 100	7	160 and under 170	117
100 and under 110	8	170 and under 180	89
110 and under 120	41	180 and under 190	57
120 and under 130	103	190 and under 200	39
130 and under 140	141	200 and under 210	10
140 and under 150	203	210 and over	14
150 and under 160	171		

(i) Calculate, or find diagrammatically, the value of the modal mass.
(ii) If we make the assumptions that the first group ranges from 80 to under 100 g and the last group from 210 to under 240 g, calculate the arithmetic mean mass and the standard deviation of the distribution.
(AEB)

5 The area of each living room in houses on an estate was calculated correct to the nearest 10 square decimetres and the areas classified as in Column 1. The cumulative frequency is given in Column 2.

Column 1 Area of room (10 square decimetres)	Column 2 Cumulative frequency
20 and under 40	3
40 and under 60	18
60 and under 80	35
80 and under 100	70
100 and under 120	100
120 and under 140	136
140 and under 160	173
160 and under 180	188
180 and under 200	210
200 and under 260	230

Calculate the (i) frequency of each class, (ii) arithmetic mean of the distribution, (iii) standard deviation of the distribution. (AEB)

6 A hundred steel tubes of internal diameter 20 cm are ordered from a workshop. On delivery they are measured to the nearest hundredth of a cm, the result being shown in the table.

Length (cm)	19·98	19·99	20·00	20·01	20·02	20·03
Number	5	22	33	20	18	2

Calculate the mean length and the standard deviation from the mean. What is measured by the standard deviation? (L)

7   The lengths of 250 articles were measured to the nearest cm and the distribution was as follows:

Length (cm)	56	57	58	59	60	61	62	63
Number	8	27	58	71	55	24	5	2

Draw the cumulative frequency curve. Read from your graph the median and quartiles, and calculate the quartile deviation.   (L)

8   Calculate the mean, the mean deviation and the standard deviation of the distribution of lengths in the previous question.   (L)

9   In a certain examination the maximum mark obtainable was 80 and the marks of three candidates A, B and C were 32, 48 and 76, respectively. The arithmetic mean of all the marks was 45 and their standard deviation was 20. If the marks are expressed as percentages, what do the arithmetic mean and the standard deviation become?

If the marks are adjusted, without their relative values being altered, so that the arithmetic mean becomes 50 and the standard deviation becomes 24, find the adjusted marks of A, B and C.   (L)

10   The table shows the intelligence quotient (I.Q.) of 100 pupils at a certain school. Calculate the (i) mean, (ii) mean deviation, (iii) standard deviation.

I.Q.	55–	65–	75–	85–	95–	105–	115–	125–	135–
No. of pupils	1	3	7	20	32	25	10	1	1

Note: 55– means 'from 55·0 to 64·9 inclusive', each I.Q. being given correct to one decimal place.   (C)

11   If the frequency of a measurement x is f and x_0 is any number, prove that the mean value M of the measurements is equal to

$$N^{-1}\Sigma f(x-x_0)+x_0$$

where $N = \Sigma f$ and the summations include all non-zero values of f. Prove also the standard deviation σ of the measurements is given by

$$\sigma^2 = N^{-1}\Sigma f(x-x_0)^2-(x_0-M)^2$$

The frequency f with which x α-particles are emitted from a radioactive specimen in a given time is shown in the following table:

x	0	1	2	3	4	5	6	7	8	9	10	11	12
f	57	203	383	525	532	408	273	139	45	27	10	4	2

Using $x_0 = 4$ as a trial mean, or otherwise, calculate the mean and standard deviation of the observed values of x. (C)

12 Define the arithmetic mean $\bar{x}$ and the variance σ^2 of a set of n numbers $x_1, x_2, \ldots, x_n$. Prove that, if a is an arbitrary number and

$$\mu_2(a) = \frac{1}{n} \sum_{r=1}^{n} (x_r - a)^2$$

then $\sigma^2 = \mu_2(a) - (\bar{x} - a)^2$

Ten rounds are fired from a gun at the same elevation, and their ranges in metres, to the nearest ten metres, are

7250 7280 7210 7290 7320 7190 7280 7250 7210 7270

Calculate the mean range, the standard deviation, and the mean deviation.
(O and C)

13 The standard deviations of two samples, each consisting of n observations, are s_1 and s_2 and their means are m_1 and m_2, respectively. Show that the standard deviation s of the combined sample of $2n$ observations is given by

$$s^2 = \tfrac{1}{2}(s_1{}^2 + s_2{}^2) + \tfrac{1}{4}(m_1 - m_2)^2$$

Six measurements are made of the percentage of acid in a chemical product and these measurements have mean 5·25 and standard deviation 0·25. A further six measurements give the following percentages of acid:

4·96 5·15 4·80 4·90 5·15 5·20

Calculate the mean and standard deviation of the twelve measurements.
(O and C)

14 If $\bar{x}$ is the arithmetic mean of the n numbers $x_1, x_2, \ldots, x_n$, prove that

$$\Sigma(x - \bar{x})^2 = \Sigma x^2 - n\bar{x}^2$$

Independent random samples are taken as follows:

20 from a population having mean 2 and standard deviation 1,
30 from a population having mean 5 and standard deviation 2, and
40 from a population having mean 4 and standard deviation 3.

Find the expectations of the mean and variance of the whole 90, regarded as one sample. (AEB)

CHAPTER 5

Regression Lines by the Method of Least Squares

Relationship between Variables

Often if corresponding values of two variables are plotted, the resulting graph suggests an algebraical relation between the variables. The diagram is known as a scatter diagram and it is advantageous to be able to express the relation in mathematical form. In *Elementary Statistics* it is shown how to draw a line of best fit, or regression line, in a linear relation, and why it is necessary in some cases to have two lines of regression unless the relation is very close. If the relation can be expressed by a straight line, there is said to be linear correlation between the variables.

The relation between variables can vary from a simple linear one to a complicated curve, but since the complicated curve can often be reduced to a straight line we will first show how to find the equation of the 'straight line of best fit'. The advantage of calculating a line of best fit over the graphical method is that no choice is left to the individual and everyone obtains the same result.

Equation of the Line of Best Fit

In the chapter on dispersion we showed that the sum of the squares of deviations from the arithmetic mean was a minimum. For this reason it is assumed that the mean of a population of all possible observed values of a measurable quantity is the true value of the quantity, and that the deviations from this value are due to random experimental errors. In practice it is impossible to obtain all the possible observed values of a quantity, and the best estimate of the true value is the mean of a sample, and the mean of the sample is the observed value from which the sum of the squares of the deviations of individual items is least. We make use of this to find the equation of the line of regression. Suppose we have a series of points $P_1, P_2, \ldots P_n$, (Diagram 5.1) whose coordinates are $(x_1 y_1)$, $(x_2 y_2) \ldots (x_n y_n)$ and that AB is a straight line whose equation is $y = ax + b$. This gives y in terms of x and is the equation of the line of regression of y on x.

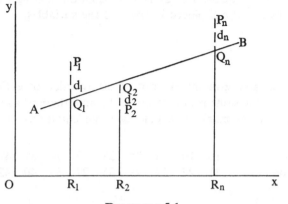

DIAGRAM 5.1

We assume that the line AB best represents the points $P_1, P_2, \ldots P_n$ when the sum of the squares of the deviations of the points from the line is a minimum. That is AB is the average line when

$$(P_1 Q_1)^2 + (P_2 Q_2)^2 + \ldots + (P_n Q_n)^2$$

is a minimum. Since $OR_1 = x_1$ we have by substituting in

$$y = ax + b$$
$$Q_1 R_1 = ax_1 + b$$

Therefore the deviation of P_1 from AB is

$$P_1 Q_1 = P_1 R_1 - Q_1 R_1 = y_1 - ax_1 - b$$

and the sum of the squares of such deviations is

$$\Sigma(y_r - ax_r - b)^2$$

Students familiar with elementary calculus will know that to find the minimum value of this expression we differentiate it first with respect to a keeping b constant, and then with respect to b keeping a constant and equating the two results to zero. (Mathematical Note 5.1.) We then have

$$\Sigma y_r = a\Sigma x_r + nb \qquad (5.1)$$
$$\Sigma x_r y_r = a\Sigma x_r^2 + b\Sigma x_r \qquad (5.2)$$

From these equations we can calculate the values of a and b. These equations are quite easy to remember because Equation (5.1) is the equation of the straight line $y = ax + b$ with a Σ in front of each variable and n in front of b, while Equation (5.2) is the equation $y = ax + b$ multiplied throughout by x and a Σ placed in front of the variables.

EXAMPLE

The following table gives marks out of 50 awarded in a French and a German test to the same group of boys. Assume there is a linear relation between the sets of marks and calculate the equations of the lines of regression.

French (x)	10	10	18	25	28	33	34	39	42	43
German (y)	11	22	22	19	35	27	33	40	42	47

SOLUTION

French Marks (x)	German Marks (y)	x^2	xy	y^2
10	11	100	110	121
10	22	100	220	484
18	22	324	396	484
25	19	625	475	361
28	35	784	980	1225
33	27	1089	891	729
34	33	1156	1122	1089
39	40	1521	1560	1600
42	42	1764	1764	1764
43	47	1849	2021	2209
$\Sigma x = 282$	$\Sigma y = 298$	$\Sigma x^2 = 9312$	$\Sigma xy = 9539$	$\Sigma y^2 = 10066$

Substituting in Equations (5.1) and (5.2) we have,

$$298 = 282a + 10b$$
$$9539 = 9312a + 282b$$

Solving for a and b we have the equation of the line of regression of y on x as

$$y = 0 \cdot 84x + 6 \cdot 25$$

This is the equation giving the German marks in terms of the French marks, and is the equation that would be used to find the probable

German mark obtained by a pupil receiving x French marks. To find the equation of the French marks in terms of the German marks we must make the German marks, or y, the independent variable and the French marks, or x, the dependent variable. That is, the graph must be turned through 90°. The equations then become,

$$\Sigma x_r = a\Sigma y_r + nb$$
and $$\Sigma x_r y_r = a\Sigma y_r{}^2 + b\Sigma y_r$$

where a and b will have different values from before. Substituting the appropriate values, we have

$$282 = 298a + 10b$$
$$9539 = 10066a + 298b$$

which gives the regression line of x on y as

$$x = 0{\cdot}96y - 0{\cdot}34$$

and this is the line giving the probable French mark, x, obtained by a pupil receiving y German marks. The two lines are shown in Diagram 5.2.

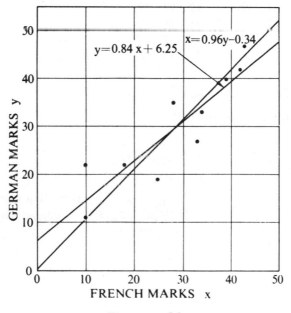

DIAGRAM 5.2

The Regression Lines and the Means

If we divide Equation (5.1) by n, the total number of items, we have

$$\frac{\Sigma y_r}{n} = \frac{a\Sigma x_r}{n} + b$$

Now $\Sigma y_r/n$ and $\Sigma x_r/n$ are the arithmetic means of the y and x variables; therefore

$$\bar{y} = a\bar{x} + b$$

That is, the regression lines pass through the means of the distributions. This property is used as an aid to drawing the regression lines.

Exercise 5.1

Calculate the equations of the lines of regression for the following distributions:

(a) x 12 30 36 24 30 34 30 38
     y 32 30 44 44 47 58 61 70
(b) x 12 24 30 34 47 58 68
     y 32 44 47 58 73 72 88
(c) x 39 61 49 64 42 72 52 57
     y 44 62 54 70 46 76 60 64
(d) x 32 51 47 35 55 41 63 72 60 45
     y 37 53 50 38 50 45 64 70 70 50

Alternative Forms of the Equations

If we solve for a in Equations (5.1) and (5.2) we have

$$a = \frac{n\Sigma xy - \Sigma x \Sigma y}{n\Sigma x^2 - (\Sigma x)^2}$$

$$= \frac{\dfrac{\Sigma xy}{n} - \dfrac{\Sigma x}{n} \times \dfrac{\Sigma y}{n}}{\dfrac{\Sigma x^2}{n} - \left(\dfrac{\Sigma x}{n}\right)^2} \quad \text{(Dividing numerator and denominator by } n)$$

$$= \frac{\dfrac{\Sigma xy}{n} - \bar{x}\bar{y}}{\dfrac{\Sigma x^2}{n} - \bar{x}^2}$$

The variance of the x's is $\Sigma(x_r - \bar{x})^2/n$ and if we represent this by $s_x{}^2$ since we cannot deal with all the possible observed points, we are dealing with a sample, and it is usual to denote the variance of a sample by s^2 and to retain σ^2 for the variance of the whole population. (This s is not to be confused with the s in the relation

$$s^2 = \sigma^2 + b^2$$

any more than the b in

$$y = ax + b$$

is to be confused with the b in this relation.) We have

$$s_x{}^2 = \frac{\Sigma(x - \bar{x})^2}{n}$$

$$= \frac{\Sigma x^2}{n} - 2\bar{x}\frac{\Sigma x}{n} + \frac{n\bar{x}^2}{n}$$

$$= \frac{\Sigma x^2}{n} - 2\bar{x}^2 + \bar{x}^2$$

$$= \frac{\Sigma x^2}{n} - \bar{x}^2$$

If the covariance of the x's and y's is defined as

$$s_{xy} = \frac{\Sigma(x - \bar{x})(y - \bar{y})}{n}$$

then $$s_{xy} = \frac{\Sigma xy}{n} - \frac{\bar{y}\Sigma x}{n} - \frac{\bar{x}\Sigma y}{n} + \frac{n\bar{x}\bar{y}}{n} \quad \text{(Multiplying the brackets)}$$

$$= \frac{\Sigma xy}{n} - \bar{x}\bar{y} \quad (\Sigma x = n\bar{x}, \Sigma y = n\bar{y})$$

The value of a can now be written as

$$a = s_{xy}/s_x{}^2$$

This is the slope of the line, or the regression coefficient, and since we have proved the regression line passes through $\bar{x}$, $\bar{y}$, its equation becomes

$$y - \bar{y} = (s_{xy}/s_x{}^2)(x - \bar{x})$$

This is the equation of the line of regression of y on x. Similarly the equation of the line of regression of x on y is

$$x - \bar{x} = (s_{xy}/s_y{}^2)(y - \bar{y})$$

Simplification of the Computation

In Chapter 1 on the Σ notation we drew attention to an important example (page 7), which said that if

then

$$x = a+bX \quad \text{and} \quad y = c+dY$$
$$\bar{x} = a+b\bar{X} \quad \text{and} \quad \bar{y} = c+d\bar{Y}$$

$$\frac{s_{xy}}{s_x^2} = \frac{ds_{XY}}{bs_X^2} \qquad \frac{s_{xy}}{s_y^2} = \frac{bs_{XY}}{ds_Y^2}$$

We can now make use of these relations to simplify the computation for the equations of the regression lines by reducing the magnitude of the items.

If in the last example (page 52) we take $a = c = 30$, and $b = d = 1$ we have,

$$x = 30+X \quad \text{and} \quad y = 30+Y$$
$$\bar{x} = 30+\bar{X} \quad \text{and} \quad \bar{y} = 30+\bar{Y}$$

$$\frac{s_{xy}}{s_x^2} = \frac{s_{XY}}{s_X^2} \qquad \frac{s_{xy}}{s_y^2} = \frac{s_{XY}}{s_Y^2}$$

Arranging the computation in terms of X and Y we have

X		Y		X²	Y²	XY
−20	(10–30)	−19	(11–30)	400	361	380
−20	(10–30)	−8	and so on	400	64	160
−12	(18–30)	−8		144	64	96
−5	and so on	−11		25	121	55
−2		5		4	225	−10
3		−3		9	9	−9
4		3		16	9	12
9		10		81	100	90
12		12		144	144	144
13		17		169	289	221
−18		−2		1392	1386	1139

Therefore,

$$\bar{x} = 30+\frac{-18}{10} = 28{\cdot}2$$

$$\bar{y} = 30+\frac{-2}{10} = 29{\cdot}8$$

$$\frac{s_{xy}}{s_x^2} = \frac{s_{XY}}{s_X^2} = \frac{113{\cdot}9-(-1{\cdot}8)(-0{\cdot}2)}{139{\cdot}2-(-1{\cdot}8)^2} = 0{\cdot}835$$

Therefore the equation of the line of regression of y on x is,

$$y-\bar{y} = \frac{s_{xy}}{s_x^2}(x-\bar{x})$$

that is, $(y - 29 \cdot 8) = 0 \cdot 835(x - 28 \cdot 2)$
or $y = 0 \cdot 84x - 6 \cdot 25$

Similarly the equation of the line of regression of x on y is

$$x = 0 \cdot 96y - 0 \cdot 34$$

Exercise 5.2

1   Calculate by the last method the equations of the lines of regression for the following distributions:

(a) x 10   50   100   60   80
      y 10   40   50   40   60
(b) x 12   24   30   34   47   58   68
      y 32   44   47   57   73   72   88
(c) x 39   61   49   64   42   72   52   57
      y 44   62   54   70   46   76   60   64
(d) x 32   51   47   35   55   41   63   72   60   45
      y 37   53   50   38   50   45   64   70   70   50

The Regression Lines of More Difficult Distributions

In *Elementary Statistics* it is shown how to draw the regression lines for a bivariate distribution with variables associated more than once. We will now show how the methods explained in this chapter can be applied to calculate the lines of regression for such a distribution.

EXAMPLE

Calculate the equations of the regression lines for the following distribution.

y \ x	100	200	300	400	500	600	700	800
0	6							
50	2	5	2					
100	5	7	3	1	1			
150	2	6	1	3	1	2		
200	1	17	7	0	1			
250				2	1	1		
300					1	1		
350					1			
400						3	1	
450								1

Note. The numbers in the body of the table give the frequency of the points. Thus $x = 200$, $y = 150$ occurs 6 times, and so on.

y / Y	x / X →	100 (−3)	200 (−2)	300 (−1)	400 (0)	500 (1)	600 (2)	700 (3)	800 (4)	C_1 f_Y	C_2 $f_Y Y$	C_3 $f_Y Y^2$	C_4 $f_Y XY$
0 (−4)		12 · 6 · 72								6	−24	96	72
50 (−3)		9 · 2 · 18	6 · 5 · 30	3 · 2						9	−27	81	54
100 (−2)		6 · 5 · 30	4 · 7 · 28	2 · 3	0 · 1	−2 · 1 · −2	−4 · 2			17	−34	68	62
150 (−1)		3 · 2 · 6	2 · 6 · 12	1 · 1	0 · 3	−1 · 1 · −1	0 · 0			15	−15	15	14
200 (0)		0 · 1 · 0	0 · 17	0 · 7	0 · 0	0 · 1	0 · 0			26	0	0	0
250 (1)					0 · 0	1 · 1	2 · 1 · 2			4	4	4	3
300 (2)					0 · 2	2 · 1 · 2	4 · 1 · 4			2	4	8	6
350 (3)					0 · 2	3 · 1 · 3	6 · 0			1	3	9	3
400 (4)							8 · 0	12 · 1 · 12		4	16	64	36
450 (5)									20 · 1 · 20	1	5	25	20
R_1 f_x		16	35	13	6	6	7	1	1	Σf_Y 85	$\Sigma f_Y Y$ −68	$\Sigma f_Y Y^2$ 370	$\Sigma f_Y XY$ 270
R_2 $f_x X$		−48	−70	−13	0	6	14	3	1				
R_3 $f_x X^2$		144	140	13	0	6	28	9	4				
R_4 $f_x XY$		126	70	13	0	3	26	12	20				

Σf_x 85 $\Sigma f_x X$ −104 $\Sigma f_x X^2$ 356 $\Sigma f_x XY$ 270

METHOD

1 If possible simplify the values of x and y. Thus, if $x = 400 + 100X$ (i.e. $a = 400$, $b = 100$ in $x = a + bX$) the values of X range from -3 to 4, and if $y = 200 + 50Y$ (i.e. $c = 200$, $d = 50$ in $y = c + dY$) the values of Y range from -4 to 5.

2 In the top left-hand corner of each frequency square write the value of XY. Thus when $X = -3$ and $Y = -4$ the value of $XY = 12$ and this is written in the top left-hand corner of the frequency square containing the frequency 6. In the bottom right-hand corner of the frequency square write the value of fXY, where f is the corresponding frequency. In the example $X = -3$, $Y = 6$; and therefore $fXY = (6)(-3)(-4) = 72$. Repeat for all the frequency squares.

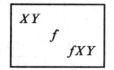

(When we deal with the variable X we denote f by f_X, and when we deal with the variable Y we denote f by f_Y.)

3 Find the sum of the frequencies f_X corresponding to the particular values of X and place these in a row below the table (R_1). Thus the number of points that have $X = -3$ are $6 + 2 + 5 + 2 + 1 = 16$.

4 Multiply each frequency in R_1 by its corresponding value of X and enter in R_2. Thus when $X = -3$ and $f_X = 16$ we have $f_X X = 16(-3) = -48$.

5 Multiply each value of $f_X X$ in R_2 by its value of X and obtain $f_X X^2$. Thus -48 is multiplied by -3 and we obtain 144. Enter these in R_3.

6 Add all the values of fXY from the bottom right-hand corners of the squares corresponding to a given value of X and obtain $f_X XY$. Enter these in R_4. Thus for $X = -3$, $f_X XY = 72 + 18 + 30 + 6 + 0 = 126$.

7 Repeat the procedures 3 to 6 for the different values of Y and thus obtain the columns C_1, C_2, C_3, and C_4.

8   Sum each row and column and obtain Σf_X, $\Sigma f_X X$, $\Sigma f_X X^2$, $\Sigma f_X XY$, Σf_Y, $\Sigma f_Y Y^2$, $\Sigma f_Y XY$. Notice $\Sigma f_X = \Sigma f_Y$ and $\Sigma f_X XY = \Sigma f_Y XY$, as is to be expected. Use these as a check on accuracy.

9   Proceed with the calculation of the regression lines as follows:

$$X = \frac{\Sigma f_X X}{\Sigma f_X} = \frac{-104}{85} = -1\cdot224$$

$$Y = \frac{\Sigma f_Y Y}{\Sigma f_Y} = \frac{-68}{85} = -0\cdot8$$

$$\bar{x} = 400 + 100(-1\cdot224) = 277\cdot6$$
$$\bar{y} = 200 + 50(-0\cdot8) = 160$$

$$s_{XY} = \frac{\Sigma f XY}{n} - \left(\frac{\Sigma f_X X}{n}\right)\left(\frac{\Sigma f_Y Y}{n}\right) = \frac{270}{85} - \left(\frac{-104}{85}\right)\left(\frac{-68}{85}\right) = 2\cdot20$$

$$s_X{}^2 = \frac{\Sigma f_X X^2}{n} - \left(\frac{\Sigma f_X X}{n}\right)^2 = \frac{356}{85} - \left(\frac{-104}{85}\right)^2 = 2\cdot69$$

$$s_Y{}^2 = \frac{\Sigma f_Y Y^2}{n} - \left(\frac{\Sigma f_Y Y}{n}\right)^2 = \frac{370}{85} - \left(\frac{-68}{85}\right)^2 = 3\cdot71$$

Therefore, $\dfrac{s_{xy}}{s_x{}^2} = \dfrac{d s_{XY}}{b s_X{}^2} = \dfrac{50 \times 2\cdot20}{100 \times 2\cdot69} = 0\cdot41$ $(b = 100, d = 50)$

The equation of the regression line of y on x is

$$y - \bar{y} = \frac{s_{xy}}{s_x{}^2}(x - \bar{x})$$

or $$y - 160 = 0\cdot41(x - 277\cdot647)$$
or $$y = 0\cdot41x + 46\cdot37$$

Similarly, $\dfrac{s_{xy}}{s_y{}^2} = \dfrac{b s_{XY}}{d s_Y{}^2} = \dfrac{100 \times 2\cdot20}{50 \times 3\cdot71} = 1\cdot18$

and the equation of the regression line of x on y is

$$x - \bar{x} = \frac{s_{xy}}{s_y{}^2}(y - \bar{y})$$

or $$x - 277\cdot647 = 1\cdot18(y - 160)$$
or $$x = 1\cdot18y - 88\cdot29$$

Trend Lines

The method of 'least squares' may be used to find the equation of the trend line of a given set of data.

EXAMPLE

Find the equation of the trend line for the data of the following table:

Annual Mean of the Daily Mean Air Temperature at Sea-level, 1952–1961

Year	1952	1953	1954	1955	1956	1957
Annual Mean Temp. (°C)	9·7	10·4	9·8	9·8	9·4	10·6

	1958	1959	1960	1961
	9·9	10·9	10·2	10·5

SOLUTION

Year	x	$y =$ temp	x^2	xy
1952	−9	9·7	81	−87·3
1953	−7	10·4	49	−72·8
1954	−5	9·8	25	−49·0
1955	−3	9·8	9	−29·4
1956	−1	9·4	1	−9·4
1057	1	10·6	1	10·6
1958	3	9·9	9	29·7
1059	5	10·9	25	54·5
1960	7	10·2	49	71·4
1961	9	10·5	81	94·5
	0	101·2	320	12·8

Note. Since we have an even number of years it simplifies the calculation if we take six months as a unit and the origin as midway between 1956 and 1957. This makes Σx equal to zero. In the calculation of constants to find the curve of best fit this choice of unit and origin are very helpful because the sums of the odd powers of x are all zero.

Substituting in Equations (5.1) and (5.2) we have

$$101·2 = 10a_0 + 0$$
$$12·8 = 0 + 320a_1$$

Solving $a_0 = 10·12$, $a_1 = 0·04$, and the equation becomes

$$y = 10·12 + 0·04x$$

which shows the annual mean of the daily mean temperature tended to increase over the 10 years by approximately 0·72°C. [The difference in the value of y as x changes from -9 to $+9$.]

Exercise 5.3

1 Calculate the equations giving the most probable value of y in terms of x, and x in terms of y for the following distributions.

(i)

y \ x	1	2	3	4	5
1				1	
2		1	1	2	
3	1	2	1		
4	1	2			
5	2	1			

(ii)

y \ x	1	2	3	4	5	6	7	8	9
9									1
8								1	
7							1	2	
6						1	2	1	
5						2	3	2	
4				1	2	2	1		
3			1	2	2	1			
2		1	2	1	1				
1	2	1							

2 The following table gives the corresponding heights and masses of schoolgirls at the same age (fictitious figures). Calculate the equations of the regression lines.

Height and Mass of Schoolgirls

Mass (kg) (Central values)	Height in cm (Central values)										
	100	102·5	105	107·5	110	112·5	115	117·5	120	122·5	125
35·5	1										
37·5		2	3	2	1						
39·5	1	1	3	5	3		1				
41·5	1	1	3	4	4	6	1		1		
43·5		1	1	4	10	9	4				
45·5		1	1	3	5	10	6	3			
47·5	1		2	3	6	9	12	6	2	1	
49·5			1	1	1	4	6	6	3		
51·5					1	2	5	5	5	1	
53·5			1			1	3	4	6	2	
55·5						1	2	3	3	1
57·5							1	1	1	1
59·5							1		1	1

3 The diameters (d) of fibres 2 cm long were measured in terms of micrometres (= one millionth of a metre), and the breaking loads (b) in terms of newtons. The frequency table for $\log_{10} b$ and $\log_{10} d$ (fictitious

figures) is given below. Calculate an equation giving the most probable value of b in terms of d.

Breaking Weights and Diameters of Fibres

$\log_{10} b$	1·15–1·20	1·20–1·25	1·25–1·30	1·30–1·35	1·35–1·40	1·40–1·45	1·45–1·50	1·50–1·55	1·55–1·60	1·60–1·65	1·65–1·70
−0·10–0·00	1	1									
0·00–0·10	1	6	3								
0·10–0·20	3	6	1								
0·20–0·30	8	13	3	1	1						
0·30–0·40	2	5	5	2	2						
0·40–0·50				5	8						
0·50–0·60				3	7	2					
0·60–0·70					5	13	1	1			
0·70–0·80					3	6	5	5			
0·80–0·90						3	9	3	2		
0·90–1·00							2	12	1	1	
1·00–1·10							2	10	4	1	
1·10–1·20								1	3	1	1

4 The frequency table gives the lengths and breadths of skulls in milli-metres (fictitious figures). Calculate the equations of the regression lines.

Length and Breadth of Skulls

Length (mm) (Central values)	Breadth (mm) (Central values)										
	133	136	139	142	145	148	151	154	157	160	163
162				1	2						
165	2		1	1		2	1				
168			2	2	2	1		2			
171	2	2	2	7	9	4	5	3			
174		1	1	10	15	11	9	2	1		
177		1	4	13	12	18	7	4	1		
180				4	13	17	14	8	3	2	1
183			3	1	13	14	5	6	2	1	
186				2	6	4	4	3			1
189				1	3	2	1	2	1	1	
192				1	1	3	1	2			
195							1	1			

5 The following table gives the quarterly sales (in thousands) of a certain model of car over a period of four years.

Year	Jan.–March	April–June	July–Sept.	Oct.–Dec.
1964	25	26	23	18
1965	21	21	18	12
1966	16	16	16	9
1967	13	12	10	6

Take the end of March 1966 as the origin for the time axis and calculate the equation of the trend line.

6   The results of tests in Mathematics and Physics are given in the table below. Calculate the equations of the regression lines and distinguish between them.

Table

Physics Marks	1	2	3	4	5	6	7	8	9	10
10									1	1
9									3	3
8							2	3	4	6
7					1	2	3	3	8	3
6			1	1	3	2	8	4	4	1
5		1	3	8	9	8	3	3	4	1
4		1	6	10	11	15	2	2	1	
3	1	5	8	4	5	6	1			
2	2	3	2	2	2	1				
1	1		2							

Mathematics marks

A pupil received 7 marks in Mathematics but was absent for the Physics test. What is his most probable mark in Physics?

A different pupil received 4 marks in Physics but was absent for the Mathematics test. What is his most probable mark in Mathematics? (AEB)

7   A set of pupils sat a mock O level examination in mathematics in February and the official examination in July. The marks obtained by the candidates in both examinations were as follows:

Mock marks	8	12	21	34	34	42	43	53	54	58	70	83
Official marks	20	5	15	28	41	42	50	42	54	62	72	85

Calculate the equations of the lines of regression.

A pupil obtained 50 marks in the mock examination and was absent from the official examination. What would have been his probable mark for the official examination? (AEB)

8   Three variables, x, y and z are observed together on seven occasions and the following values obtained:

Occasion	1	2	3	4	5	6	7	Total	Mean
x	4	8	11	9	14	16	12	74	10·57
y	3	9	8	7	14	12	11	64	9·14
z	17	12	4	2	5	1	−4	37	5·29

Show that the corrected sum of squares (C.S.S.) of x $\left[\text{i.e. } \sum_{7}^{1}(x-\bar{x})^2\right]$ and the corrected sum of products (C.S.P.) of y with z $\left[\text{i.e. } \sum_{7}^{1}(y_i-\bar{y})(z_i-\bar{z})\right]$

are approximately 95·71 and −95·29. Given that the C.S.S.'s of y and z are 78·86 and 299·43, and that the C.S.P.'s of x with y and x with z are 78·43 and −127·14, obtain the regression equations of x on y, x on z and y on z. Find the values of x predicted by these equations for (a) $z = 5$, (b) $z = 15$, and also the values of x predicted when (c) $y = 9·23$, (d) $y = 6·05$. Comment on the results of these calculations.   (C)

9   The variables x, y are observed together n times, giving values x_1, y_1; x_2, y_2; ...; x_n, y_n. The sums of squares and products are calculated from the formulae

$$S_{xx} = \sum_{1}^{n} (x_i - \bar{x})^2,$$

$$S_{xy} = \sum_{1}^{n} (x_i - \bar{x})(y_i - \bar{y}),$$

$$S_{yy} = \sum_{1}^{n} (y_i - \bar{y})^2$$

The average relationship between x and y may be assumed to be a straight line of unknown gradient β through the point $(\bar{x}, \bar{y})$, so that the value, Y_i, predicted for any given x_i is

$$Y_i = \bar{y} + \beta(x_i - \bar{x})$$

Show that the value of β which minimizes the expression

$$R = \sum_{1}^{n} (y_i - Y_i)^2$$

is S_{xy}/S_{xx}, and find the minimum value of R.
   The following values were obtained from 12 pairs of observations:

$$\bar{x} = 4, \quad \bar{y} = 5; \quad S_{xx} = 10, \quad S_{xy} = -15, \quad S_{yy} = 25$$

Plot the line which minimizes R.   (C)

10   The number x of vehicles per kilometre of road in Great Britain and the maintenance cost y (million pounds) of roads are given in the following table for the years 1948 to 1957 inclusive:

| x | 20·3 | 22·3 | 24·0 | 25·0 | 26·4 | 28·4 | 30·9 | 34·1 | 36·6 | 39·9 |
| y | 26·6 | 23·5 | 27·0 | 26·3 | 32·6 | 35·7 | 36·5 | 39·8 | 42·2 | 43·2 |

Calculate
   (i) the mean and variance of x,
   (ii) the covariance of x and y,
   (iii) the equation of the line of regression of y on x.

11   The mark y obtained out of a possible 70 by each of 200 children is paired with the child's age x months. The results are shown in the following grouped frequency table:

Mark	Age in months x						
y	120	123	126	129	132	135	138
0–9	1	5	3				
10–19		4	11	2			
20–29		5	13	16	3	2	
30–39		2	4	21	13	7	3
40–49		1	3	12	16	8	4
50–59				4	6	11	3
60–69				3	2	4	8

Thus 13 children aged 125 to 127 months obtained marks between 20 and 29 inclusive.

For each central value of x calculate the mean mark obtained. Plot these mean marks against x on a graph and sketch in a regression line of y on x. Obtain the equation of the regression line in the form $y = mx + c$.

If it were desired to adjust the marks obtained so as to make allowance for age, how many marks should be added to those obtained by a child aged 126 months and how many should be subtracted from those obtained by a child aged 134 months?   (O and C)

12   The following table gives the quarterly sales (in thousands) of a certain model of car over a period of three years.

	Quarter			
Year	1st	2nd	3rd	4th
1967	23	23	20	14
1968	18	18	18	11
1969	15	14	12	8

Assuming the trend line is straight, calculate its equation. Estimate the number of cars sold in the first quarter of 1970.   (AEB)

13   American tourists visiting Great Britain

Year	Year with 1953 as origin	No. of tourists (thousands)	Amount they spent (£M)
	t	x	y
1953	0	185	36·1
1954	1	203	38·8
1955	2	240	42·5
1956	3	255	46·3
1957	4	266	47·7
1958	5	325	57·0
1959	6	360	64·0

Use the above data to plot graphs of (i) x against t and (ii) y against t and in each case draw by eye the straight line which best fits the points plotted. Obtain the equations of your straight lines in the form $x = m_1 t + c_1$, $y = m_2 t + c_2$.

Extend the table to show the estimated number of tourists and the amount which they will spend in each of the years 1960 and 1961. Discuss the reliability of these estimates. (JMB)

Mathematical Note 5.1. To derive Equations (5.1) and (5.2)

$$\Sigma d^2 = \Sigma(y_r - a x_r - b)^2$$

Let
$$S = \Sigma d^2, \text{ then}$$
$$S = \Sigma(y_r - a x_r - b)^2$$

Then for S to be a minimum

$$\frac{\partial S}{\partial b} = 0 \quad \text{and} \quad \frac{\partial S}{\partial a} = 0$$

$$\frac{\partial S}{\partial b} = -2\Sigma(y_r - a x_r - b) = 0$$

Therefore
$$\Sigma y_r = a\Sigma x_r + nb \tag{5.1}$$

$$\frac{\partial S}{\partial a} = -2\Sigma(y_r - a x_r - b)x_r = 0$$

Therefore
$$\Sigma x_r y_r = a\Sigma x_r^2 + b\Sigma x_r \tag{5.2}$$

CHAPTER 6

Equations Reducible to the Linear Form

Non-linear Relations

There are many variables which have a non-linear relation but which by a rearrangement of the variables may be reduced to a linear form. The method is best illustrated by examples.

EXAMPLE

The table gives corresponding values of the resistance R newtons per kg of a body and its velocity v metre/s. Calculate the equation for R in terms of v.

v m/s	10	20	30	35	40	45	50	55	60
R N per kg	9·4	10·4	10·6	11·7	12·9	13·7	14·1	15·7	16·7

SOLUTION

Plotting R against v, the shape of the curve (Diagram 6.1) suggests a parabolic relation of the form

$$R = a_0 + a_1 v^2$$

Test this by letting $V = v^2$ and plotting R against V. This should give us a straight line

$$R = a_0 + a_1 V$$

v	10	20	30	35	40	45	50	55	60
V	100	400	900	1225	1600	2025	2500	3025	3600

From Diagram 6.2 we see the points lie very close to a straight line.

The value of a_0 can be read from the graph; it is the value of R when v equals zero, that is 9·16. The easiest way to calculate a_1 is to substitute. Thus when $V = 2500$, $R = 14·5$.

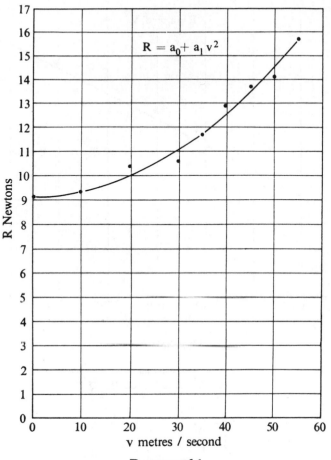

$$R = a_0 + a_1 v^2$$

R Newtons

v metres / second

DIAGRAM 6.1

Therefore
$$14.5 = 9.16 + 2500a_1$$
$$a_1 = 0.0021$$

Therefore
$$R = 9.16 + 0.0021V$$

In terms of the original variables,

$$R = 9.16 + 0.0021v^2$$

An alternative method of solution is given in the paragraph dealing with the 'parabola of best fit' on page 74.

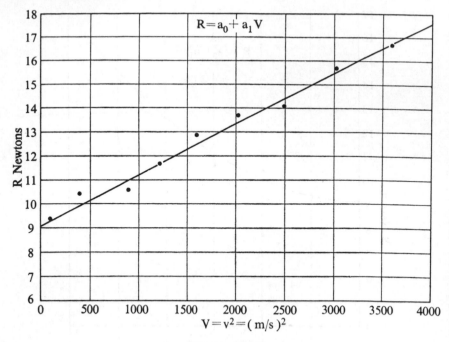

$$R = a_0 + a_1 V$$

DIAGRAM 6.2

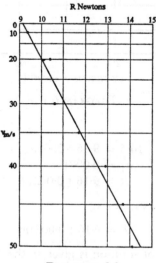

DIAGRAM 6.3

Square Law Graph Paper

If R had been plotted against v on square law graph paper we would have obtained a straight line without the necessity of squaring v. This may appear a small saving in time and work in this example but is a big saving if we have awkward numbers to deal with. Square law graph paper has one scale giving the squares of the variable. Thus if 0 to 1 is 0·5 cm, then 0 to 2 is 2 cm, i.e. 4 times 0·5; 0 to 3 is 4·5 cm, i.e. 9 times 0·5, and so on. (See Diagram 6.3.)

EXAMPLE

The table gives corresponding values of two variables x and y. Find an equation that expresses satisfactorily the variable x in terms of the variable y.

x	1	2	3	4	5	6
y	132·5	185	240	340	430	600

SOLUTION 1

Plot x against y. The shape of the curve within the limits of the magnitude of the variables suggests a parabola. If y is plotted against x^2, or against x on square law graph paper, the points lie fairly close to a straight line. Calculate the equation of this straight line by the method of the previous example. The equation is

$$y = 124 \cdot 83 + 12 \cdot 95 x^2$$

SOLUTION 2

Sometimes there may be more than one possible solution. Thus if we plot x against $\log y$, or use semi-logarithmic paper and plot x against y where x is on the linear scale and y on the logarithmic scale (Diagram 6.4), we again approximate to a straight line, which suggests a relation

$$\log y = a_0 + a_1 x$$

If we write $a_0 = \log a$ and $a_1 = \log b$ the equation becomes

$$\log y = \log a + x \log b$$

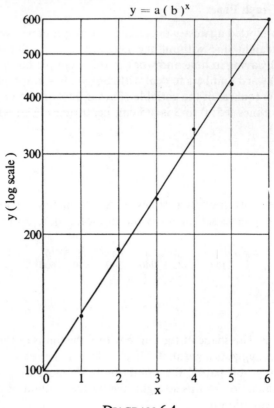

DIAGRAM 6.4

To find the constants *a* and *b* proceed as follows:

y	$Y = \log y$	x	x^2	xY
132·5	2·1222	1	1	2·1222
185	2·2672	2	4	4·5344
240	2·3802	3	9	7·1406
340	2·5315	4	16	10·1260
430	2·6335	5	25	13·1675
600	2·7782	6	36	16·6692
Totals	14·7128	21	91	53·7599

Substituting in Equations (5.1) and (5.2) we have

$$14{\cdot}7128 = 6a_0 + 21a_1$$
$$53{\cdot}7599 = 21a_0 + 91a_1$$

and solving
$$a_0 = 1{\cdot}999 \simeq 2$$
$$a_1 = 0{\cdot}1295$$

Therefore $\quad \log a = 2 \quad$ and $\quad \log b = 0{\cdot}1295$

or $\qquad\qquad a = 100 \quad$ and $\quad b = 1{\cdot}35$

Therefore $\qquad \log y = \log 100 + x \log 1{\cdot}35$

or $\qquad\qquad y = 100(1{\cdot}35)^x$

The value of the constants a and b can be found from Diagram 6.4, for when $x = 0$, $y = 100$, and when $x = 6$, $y \simeq 600$. Therefore $a = 100$ and $600 = 100b^6$, or $b = 1{\cdot}35$. Both these equations describe well the relation between x and y. How to decide which is the better one will be explained in Chapter 13.

EXAMPLE

The following table gives the population of England and Wales at each census from 1811 to 1911. Find an equation that closely expresses the relation between the population and the year.

Year	1811	1821	1831	1841	1851	1861
Population (10^6)	10·2	12·0	13·9	15·9	17·9	20·1
	1871	1881	1891	1901	1911	
	22·7	26·0	29·0	32·5	36·1	

SOLUTION

There is evidence to suggest (Mathematical Note 6.1) that if the environment remains fairly constant the rate of growth of a population can be expressed as

$$y = ae^{bx}$$

where a and b are constants and e is the exponential constant (see Chapter 11).

Taking logarithms to the base e we have

$$\log_e y = \log_e a + bx$$

If $Y = \log_e y$ and $a_0 = \log_e a$ we can write the equation as

$$Y = a_0 + bx$$

which represents a straight line.

To simplify the calculation of the constants take x in units of 10 years and the origin at 1861. Tabulating the work we have:

Year	x	Population y	$\text{Log}_e y$ Y	x^2	xY
1811	−5	10·2	2·3223	25	−11·6115
1821	−4	12·0	2·4848	16	−9·9392
1831	−3	13·9	2·6318	9	−7·8954
1841	−2	15·9	2·7662	4	−5·5324
1851	−1	17·9	2·8847	1	−2·8847
1861	0	20·1	3·0006	0	0
1871	1	22·7	3·1223	1	3·1223
1881	2	26·0	3·2580	4	6·5160
1891	3	29·0	3·3672	9	10·1016
1901	4	32·5	3·4812	16	13·9248
1911	5	36·1	3·5862	25	17·9310
Total	0		32·9050	110	13·7325

Substituting in Equations (5.1) and (5.2), we have

$$32\cdot9050 = 11a_0 + 0$$
$$13\cdot7325 = 0 + 110b$$

Therefore
$$a_0 = 2\cdot9914$$
$$b = 0\cdot1248$$

Therefore $\quad \log_e a = 2\cdot9914 \quad \text{or} \quad a = 19\cdot92$

Therefore $\quad y = 19\cdot92e^{0\cdot125x}$

or $\quad y = 20e^{(x/8)}$

If we put $x = 10$ to calculate the population in 1961 we get $y \simeq 70$ millions. In 1961 the population of England and Wales was about 46 millions, showing there must have been a drop in the birth rate after 1911. The example illustrates the danger of making long-term forecasts.

Parabolic Equations

The method of 'least squares' used to find the constants for a straight line can be extended to find constants for parabolic curves. Thus the parabola which approximates to a given set of points $(x_1, y_1)(x_2, y_2) \ldots (x_n, y_n)$ will have the equation

$$y = a_0 + a_1 x + a_2 x^2$$

where a_0, a_1, and a_2 are constants to be determined by solving the equations

$$\Sigma y = a_0 n + a_1 \Sigma x + a_2 \Sigma x^2$$
$$\Sigma xy = a_0 \Sigma x + a_1 \Sigma x^2 + a_2 \Sigma x^3 \qquad (6.1)$$
$$\Sigma x^2 y = a_0 \Sigma x^2 + a_1 \Sigma x^3 + a_2 \Sigma x^4$$

These equations are derived in a similar manner to the first set of equations and the method can be extended to equations of higher order. Thus an equation of the nth order can be derived from

$$\Sigma y = a_0 n + a_1 \Sigma x + a_2 \Sigma x^2 + \ldots + a_n \Sigma x^n$$
$$\Sigma xy = a_0 \Sigma x + a_1 \Sigma x^2 + a_2 \Sigma x^3 \ldots + a_n \Sigma x^{n+1}$$
$$\vdots \qquad \vdots \qquad \vdots$$
$$\Sigma x^n y = a_0 \Sigma x^n + a_1 \Sigma x^{n+1} \ldots + a_n \Sigma x^{2n}$$

EXAMPLE

Fit a parabola of the type

$$y = a_0 + a_1 x + a_2 x^2$$

to the data of the example on page 68.

SOLUTION

To reduce the magnitude of the items involving x, write $X = x/10$. Tabulating,

y	$X = x/10$	X^2	X^3	X^4	yX	yX^2
9·4	1	1	1	1	9·4	9·4
10·4	2	4	8	16	20·8	41·6
10·6	3	9	27	81	31·8	95·4
11·7	3·5	12·25	42·875	150·0625	40·95	143·325
12·9	4	16	64	256	51·6	206·4
13·7	4·5	20·25	91·125	410·0625	61·65	277·425
14·1	5	25	125	625	70·5	352·5
15·7	5·5	30·25	166·375	915·0625	86·35	474·925
16·7	6	36	216	1296	100·2	601·2
115·2	34·5	153·75	741·375	3750·1875	473·25	2202·175

Substituting in Equations (6.1), we have

$$115\cdot2 = 9u_0 + 34\cdot5a_1 + 153\cdot75a_2$$
$$473\cdot25 = 34\cdot5a_0 + 153\cdot75a_1 + 741\cdot375a_2$$
$$2202\cdot175 = 153\cdot75a_0 + 741\cdot375a_1 + 3750\cdot1875a_2$$

and solving

$$a_0 = 9\cdot05; \qquad a_1 = 0\cdot031; \qquad a_2 = 0\cdot21$$

Therefore, the equation of the parabola is

$$y = 9.05 + 0.031X + 0.21X^2$$

Since $X = x/10$ the equation in terms of the original variables is

$$y = 9.05 + 0.0031x + 0.0021x^2$$

which agrees very closely with the first solution.

It is important to realise that the curve calculated fits the data given, and that it may diverge by appreciable amounts from observations outside the limits given. Even within the limits given the fit may tend to worsen as the limits are approached. The equation should not be used to make estimates far from the limits of the observations.

Exercise 6.1

1   Plot y against x and show that the following points are probably connected by an equation of the form

$$y = a + bx^2$$

Calculate the values of a and b. (Use square law graph paper if available.)

y	3.5	2.5	−4.5	−13.5	−27.5	−45.5
x	0	1	2	3	4	5

2   The following are corresponding values of the resistance R N per tonne of a body, and its velocity v m/s. It is generally assumed that

$$R = a + bv^2$$

Test if this is approximately so and calculate the values of a and b.

v m/s	10	15	20	25	30	35	40
R N per tonne	200	490	650	962	1400	1900	2440

3   The table gives corresponding values of the variables y and x. Test graphically if the equation

$$y = a + \frac{b}{x}$$

gives a satisfactory relation between y and x, and calculate the values of a and b.

y	90	113	178	196	210
x	22	18	11	10	9.5

4 It is thought that in the following table $y = a + b/x$. Test if this is so and calculate the probable values of a and b.

x	2	3	4	5
y	3·1	2·9	2·7	2·6

5 The following values of R and v are expected to obey a law of the form $R = a + bv^3$. Test by plotting R against v^3 and calculate the most probable values of a and b.

R	295	550	1150	1800	2300
v	7·6	10·3	13·4	16·6	18

6 Plot the following points and from the graph deduce the probable equation connecting them. Calculate the values of the constants.

(i) y	1·6	2·9	4·8	7·7	11·2	15·7
x	2	4	6	8	10	12
(ii) y	−2	−8	−20	−30	−56	
x	1	2	3	4	5	

7 Use Equation (6.1) to fit a parabola to the following points:

(i) x	1	2	3	4	5
y	2	8	12	22	30
(ii) x	1	2	3	4	5
y	4	5	12	15	28

Plot the points and the curve on the same graph and compare the fit.

8 The following table gives the World Record y seconds, to the nearest 10 seconds, for running x kilometres, as it stood on 10 September 1962:

x	1	2	3	5
y	140	300	470	810

Use the method of least squares to fit a curve of the form

$$Y = a + bx + cx^2$$

and use the results to estimate the World Records for 1500 and 10 000 metres. (AEB)

9 By the method of least squares fit a curve of the form

$$y = a + b(x + x^2)$$

to the following data:

x	−3	−2	−1	0	1	2	3
y	17	3	−1	−3	5	15	35

Give a rough estimate of the value of y when x is 1·5. (AEB)

10   The following table gives corresponding values of two variables x and y

x	0	1	1·5	2	2·5	3	3·5
y	3·2	4·8	8	11·3	15·2	20·7	27·6

Plot the values of y against the values of x^2. By the method of least squares find a suitable equation giving y in terms of x^2.   (AEB)

11   Graph the following series on semi-logarithmic graph paper. Use the graph to compare approximately the rates of increase in imports in the decades 1861–1870 and 1871–1880:

Year	1860	1861	1862	1863	1864	1865	1866	1867	1868
Imports·(£m)	211	217	226	249	275	271	295	275	295
	1869	1870	1871	1872	1873	1874	1875	1876	1877
	295	303	331	355	371	370	374	375	394
	1878	1879	1880						
	369	363	411						

(L)

12   By means of a graph, or otherwise, find the values of a and b which give an approximate fit for the curve $y = ax^b$ to the following data:

x	1·2	1·4	2·0	3·6	5·8
y	5·4	6·9	12·1	31·1	66·4

Estimate (i) the value of y when $x = 3$, (ii) the value of x when $y = 35$.   (AEB)

Mathematical Note 6.1.   Most populations have the property that their rate of growth at any moment is proportionate to the size of the population at that moment; that is, the 'birth-rate' is constant. If this is so, and y represents the population at time x, then $dy/dx = by$ where b is a constant. Solving this differential equation,

$$dy/y = bdx$$

Therefore $$\log_e y + \text{constant} = bx$$

Let the constant equal $\log_e a^{-1}$

Then $$\log_e y + \log_e a^{-1} = bx$$
or, $$\log_e ya^{-1} = bx$$
Therefore $$ya^{-1} = e^{bx}$$
or, $$y = ae^{bx}$$

CHAPTER 7

Correlation

Correlation

In the last chapter we showed how to find the equation of the line that describes how well a linear equation gives the relation between two variables, and showed that generally two lines of regression were necessary—one to calculate y given x, and one to calculate x given y. The coefficient used to measure how close the relation is between the two regression lines, that is how near they are to coinciding, is called the coefficient of correlation.

Product-moment Correlation

The line of regression of y on x is

$$y - \bar{y} = \frac{s_{xy}}{s_x^2} (x - \bar{x})$$

This can be written as

$$\frac{y - \bar{y}}{s_y} = \frac{s_{xy}}{s_x s_y} \times \frac{x - \bar{x}}{s_x} \quad \text{(Dividing both sides by } s_y\text{)}$$

Write $\dfrac{y - \bar{y}}{s_y}$ as Y; $\dfrac{x - \bar{x}}{s_x}$ as X, and $\dfrac{s_{xy}}{s_x s_y}$ as r; then the equation can be written

$$Y = rX$$

What we have done is to move the origin of the coordinates to $\bar{x}$, $\bar{y}$, and written X and Y in units of s_x and s_y, respectively. Similarly the line of regression of x on y is

$$x - \bar{x} = \frac{s_{xy}}{s_y^2} (y - \bar{y})$$

or rewriting

$$\frac{x - \bar{x}}{s_x} = \frac{s_{xy}}{s_x s_y} \frac{(y - \bar{y})}{s_y} \quad \text{(Dividing both sides by } s_x\text{)}$$

or

$$X = rY$$

Now $Y = rX$ is the equation of a straight line through the origin making an angle θ with the X-axis where $\tan \theta = r$.

$$\tan \theta = \frac{Y}{X} = r$$

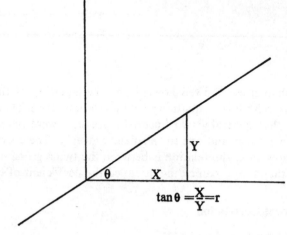

$$\tan \theta = \frac{X}{Y} = r$$

Fig. 7.1(a)

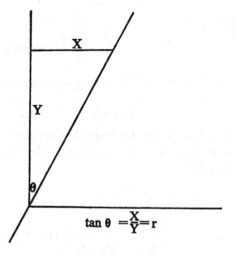

$$\tan \theta = \frac{X}{Y} = r$$

Fig. 7.1(b)

Similarly $$X = rY$$

is the equation of a straight line making the same angle θ with the Y-axis.

$$\tan \theta = \frac{X}{Y} = r$$

If the correlation between the two variables is perfect and direct, the two lines coincide, and $\theta = 45°$. Then,

$$r = \tan \theta = \tan 45° = 1$$

When there is a large amount of scatter amongst the points, the regression lines are separate and the greatest separation is when $\theta = 0$. Then,

$$r = \tan \theta = \tan 0 = 0$$

If there is perfect inverse correlation, the two lines coincide but lie in the second and fourth quadrants, and $\theta = 90°+45°$. Then,

$$r = \tan(90+45) = -1$$

r is defined as the product moment coefficient of correlation.

Calculation of r

$$r = \frac{s_{xy}}{s_x s_y}$$

where

$$s_{xy} = \frac{\Sigma xy}{n} - \frac{\Sigma x}{n} \times \frac{\Sigma y}{n}$$

$$s_x^2 = \frac{\Sigma x^2}{n} - \left(\frac{\Sigma x}{n}\right)^2$$

$$s_y^2 = \frac{\Sigma y^2}{n} - \left(\frac{\Sigma y}{n}\right)^2$$

and if

$$x_r = a + bX_r$$

and

$$y_r = c + dY_r$$

then,

$$\frac{s_{xy}}{s_x s_y} = \frac{s_{XY}}{s_X s_Y}$$ (See Mathematical Note 1.1)

EXAMPLE

The table gives the index figures for production and the price of an article over ten consecutive years. Calculate the coefficient of correlation.

Year	1	2	3	4	5	6	7	8	9	10
Production	92	96	103	108	109	108	96	103	109	103
Price	109	111	94	93	89	84	100	106	87	97

SOLUTION

(The method is explained after this table.)

Col. 1 x	Col. 2 X	Col. 3 y	Col. 4 Y	Col. 5 X²	Col. 6 Y²	Col. 7 XY
92	−11	109	12	121	144	−132
96	−7	111	14	49	196	−98
103	0	94	−3	0	9	0
108	5	93	−4	25	16	−20
109	6	89	−8	36	64	−48
108	5	84	−13	25	169	−65
96	−7	100	3	49	9	−21
103	0	106	9	0	81	0
109	6	87	−10	36	100	−60
103	0	97	0	0	0	0
1027	−3	970	0	341	788	−444

METHOD

Col. 1 shows the production figures and since the total is 1027 the mean is approximately 103.

Col. 2. Reduce the production figures by subtracting 103 from each figure in Col. 1, i.e. $x = 103 + X$.

Col. 3 shows the price figures and their mean is 97.

Col. 4. Reduce the price figures by subtracting 97 from each figure in Col. 3, i.e. $y = 97 + Y$.

Cols. 5, 6, 7 are self explanatory. Therefore

$$s_{XY} = \frac{\Sigma XY}{n} - \frac{\Sigma X}{n} \times \frac{\Sigma Y}{n}$$

$$= -44 \cdot 4 - (0 \cdot 3)(0)$$

$$= -44 \cdot 4$$

$$s_X^2 = \frac{\Sigma X^2}{n} - \left(\frac{\Sigma X}{n}\right)^2$$

$$= 34 \cdot 1 - 0 \cdot 09$$

$$= 34 \cdot 01$$

$$s_X = 5 \cdot 832$$

$$s_Y^2 = \frac{\Sigma Y^2}{n} - \left(\frac{\Sigma Y}{n}\right)^2$$

$$= 78 \cdot 8 - 0$$

$$s_Y = -8 \cdot 877$$

$$r = \frac{s_{XY}}{s_X s_Y}$$

$$= \frac{-44\cdot 4}{5\cdot 832 \times 8\cdot 877}$$

$$= 0\cdot 858$$

Variables Associated More than Once

The calculation of the coefficient of correlation for a bivariate distribution with variables associated more than once is similar to the calculation of the equations of the lines of regression of the distribution.

Thus in the example on page 57 we have $s_{XY} = 2\cdot 20$; $s_X = 1\cdot 64$; $s_Y = 1\cdot 93$; and therefore

$$r = \frac{s_{XY}}{s_X s_Y}$$

$$= \frac{2\cdot 20}{1\cdot 64 \times 1\cdot 93}$$

$$= 0\cdot 70$$

Calculations from the Equations of the Regression Lines

The equations of the regression lines are:

y on x $$y - \bar{y} = \frac{s_{xy}}{s_x^{\,2}} (x - \bar{x})$$

and x on y $$x - \bar{x} = \frac{s_{xy}}{s_y^{\,2}} (y - \bar{y})$$

Firstly, it is obvious that the solution of these equations gives us the arithmetic means of the two distributions x and y. Secondly, the coefficient of x in the first equation is $s_{xy}/s_x^{\,2}$ and of y in the second equation is $s_{xy}/s_y^{\,2}$. The product of these coefficients is $s_{xy}^{\,2}/(s_x^{\,2}s_y^{\,2})$, and by definition this equals r^2. Therefore the square root of this product equals r.

EXAMPLE

The equations of the two regression lines of a bivariate distribution are

$$y = 1\cdot 3x + 0\cdot 4$$
$$x = 0\cdot 7y - 0\cdot 1$$

Calculate the arithmetic mean of each distribution and the product-moment correlation coefficient between the two distributions.

SOLUTION

Solving the two equations we have $\bar{x} = 2$; $\bar{y} = 3$. Also,

$$s_{xy}^2/(s_x^2 s_y^2) = 1{\cdot}3 \times 0{\cdot}7 = 0{\cdot}91$$

Therefore, $r^2 = 0{\cdot}91$ and $r = \pm 0{\cdot}954$

From the equations we see that x and y increase together and we must take the positive value of r.

Therefore, $r = 0{\cdot}954$

Rank Correlation

The drawback of the product moment correlation coefficient is that the data must be definitely measured. This may not always be desirable or possible, but it may be possible to rank the items according to the degree they possess a certain property or ability. Thus it is natural to inquire if the ranks of individuals ranked according to two different qualities can be used to measure the correlation between them.

Mean Rank Correlation

Suppose we have n individuals whose ranks according to the first quality are $X_1, X_2, \ldots X_n$ and whose ranks according to the second quality are $Y_1, Y_2, \ldots Y_n$. Now $X_1, \ldots X_n$ and $Y_1, \ldots Y_n$ are the first n natural numbers $1, 2, 3, \ldots n$, but not necessarily in that order.

Let the difference between two ranks be d_k, i.e.

$$d_k = X_k - Y_k$$

A possible measure of correlation is the arithmetic mean of the absolute values of the deviations, $\dfrac{\Sigma|d|}{n}$, but this measure is inconvenient to use in analytical investigations. A better rank correlation coefficient is Spearman's Rank Correlation Coefficient, $\rho = 1 - \dfrac{6\Sigma d^2}{n(n^2-1)}$, described in *Elementary Statistics*. (For the derivation of Spearman's coefficient see Mathematical Note 7.1.)

Exercise 7.1

1 Calculate the product moment correlation coefficient for questions 1 to 4 of Exercise 5.3.

2 The following table gives the height x hundred metres above sea level, to the nearest unit of x, of twelve places in Switzerland, and the early morning temperature $y\,°C$ on the same day in August. Find the equations of the regression lines and of the correlation coefficient between x and y.

x	11	15	10	5	4	5	9	12	3	4	18	16
y	13	4	10	18	14	13	5	9	18	18	8	6

(AEB)

3 The following are estimated average unemployment percentages, 1951–61, for nine regions (Source *J.R.S.S.*, Series A, Vol. 127, Pt. 1, 1964)

Region	Males (x)	Females (y)
London, etc.	1·12	0·91
South-Western	1·60	1·61
Midland	1·04	0·99
North-Midland	0·88	1·02
East and West Ridings	1·26	1·19
North-Western	1·97	2·12
Northern	2·21	2·56
Scotland	3·22	2·92
Wales	2·58	3·79

Fit regression lines (i) for y on x, (ii) for x on y and find the coefficient of correlation between x and y. (AEB)

4 Sixteen boys were each given four problems in arithmetic and four problems in algebra. For each problem correctly answered they were given one mark, with the following result:

Candidate	A	B	C	D	E	F	G	H	I	J	K	L	M	N	O	P
Arithmetic	3	3	2	4	1	3	3	2	2	2	4	3	3	1	2	2
Algebra	3	2	2	3	2	3	4	2	1	3	4	2	3	1	3	2

Group the candidates by their arithmetic mark and find the average algebra mark for each group.

Draw the regression line of the algebra mark on the arithmetic mark, and the regression line of the arithmetic mark on the algebra mark. From the gradients of these two lines calculate the coefficient of correlation.

(L)

5 The table below shows the examination marks of eight students in Algebra and Statistics. Calculate the product-moment coefficient of

correlation between the Algebra and Statistics marks. Explain the significance of your answer

Algebra	10	24	30	35	48	59	68	70
Statistics	30	47	44	71	60	89	97	74

(AEB)

6 State briefly the meaning of the term correlation. Write down a set of five pairs of values of two variables (a) with maximum positive correlation, (b) with maximum negative correlation.

Construct a scatter diagram for the set of paired values of the variables x and y given below.

x	4	4	6	6
y	6	4	4	6

Draw the line of regression of y upon x and the line of regression of x upon y, indicating which is which.

What is the coefficient of correlation between x and y?

(L)

7 The table shows the index figure for production and for price of an article over ten consecutive years. Calculate the coefficient of correlation. Comment on your result.

Year	1	2	3	4	5	6	7	8	9	10
Production	92	96	103	108	109	108	96	103	109	103
Price	109	111	94	93	89	84	100	106	87	97

(C)

8 The frequency table given below shows the results of giving a test in English and a test in French to a class of 30 boys. Each test is marked out of 10. Calculate the coefficient of correlation and comment on your result.

English	2	3	4	5	6	7	8	9	10
10									
9							1	1	
8					1			1	
7						1			1
6		1		2		2			
5			2	1	1		2		
4				2		2	1		
3	1	2		1	2				
2			1			1			

French

(C)

9   In a study of the quality of imported bacon the following observations were obtained of the leanness of the carcase (x) and the palatability (y):

x	1·8	2·6	3·4	5·2	5·9	6·8	8·3	11·5	12·0	13·0	13·1	13·8
y	5·7	5·0	7·8	8·9	10·4	10·2	10·2	9·7	7·5	4·8	7·0	5·5

Plot the observations and calculate the correlation coefficient between x and y, using means of 7 for x and 8 for y. Comment upon the value of the correlation coefficient.   (C)

10   The individuals comprising a population are endowed with two characteristics (height and weight for example) whose numerical measures are values of the variates x and y. Show how data relating to two such paired variates can be represented on a scatter diagram, and explain what is meant by (a) the lines of linear regression of y on x and of x on y, (b) the coefficient of correlation between x and y.

Two hundred candidates sit for an examination in which there are separate papers in mathematics and mechanics. The percentage marks obtained by a candidate in these subjects are paired values of the variates x and y. When $a = 64·5$ and $b = 64·5$ are taken as trial mean values of x and y it is found that:

$$\Sigma X = +830 \qquad \Sigma X^2 = 47\,900$$
$$\Sigma Y = -840 \qquad \Sigma Y^2 = 52\,000$$
$$\Sigma XY = +25\,700$$

where $X = x-a$ and $Y = y-b$ are deviations from the trial means, and the summations include all candidates. Find:

(i) the equations of the linear regression lines of y on x and of x on y,
(ii) the coefficient of correlation between x and y.

Comment on the value of the coefficient of correlation.   (C)

11   The equations of the two regression lines of a bivariate distribution are:

$$y = -1·2x-5·6$$
$$x = -0·8y-4·6$$

Calculate the arithmetic mean of each distribution and the product-moment correlation coefficient between the two distributions.

12   The following table gives the number of road vehicles x (unit 1 000 000) in use in September of each year and the number of road casualties y (unit 100 000) in each year:

Year	Vehicles x	Casualties y
1952	4·9	2·1
1953	5·3	2·3
1954	5·8	2·4
1955	6·4	2·7
1956	6·9	2·7
1957	7·5	2·7
1958	7·9	3·0
1959	8·6	3·3

Find

(i) the covariance of x and y,

(ii) the line of regression of y on x,

(iii) the correlation coefficient between x and y.   (O and C)

13   The average price for each of eight successive months of a certain commodity in dollars in New York (y) and in pounds sterling in London (x) is given in the following table:

y \$	83	85	88	80	98	95	97	94
x £	28	29	29	27	31	32	32	32

Calculate

(i) the covariance of x and y,

(ii) the equation of the line of regression of y on x,

(iii) the correlation coefficient of x and y.   (O and C)

14   The following table shows the lunch-time temperatures on Thursday November 25th, 1965 (Source, *The Times*) and the latitude north of the Equator, to the nearest half-degree, for a number of towns:

	Lunch-time temperature °C	Latitude north of the equator
Amsterdam	4	52½
Barcelona	10	54
Belfast	2	54½
Berlin	3	52
Birmingham	10	52½
Bristol	10	51½
Brussels	7	51
Budapest	−5	47½
Edinburgh	1	56
Geneva	4	46
Helsinki	−4	60
Lisbon	7	39
London	9	51½
Madrid	4	40½

Moscow	−10	55½
New York	8	40½
Paris	8	48½
Rome	11	42
Vienna	4	48
Warsaw	−3	52

Calculate

(i) the coefficient of correlation and
(ii) the coefficient of correlation by ranks, for the two sets of figures.

(AEB)

15 Calculate the equations of the regression lines and the coefficient of correlation for the following bivariate frequency table for values of x and y:

x $\diagdown$ y	−10	−5	0	5	10	Total
0	80	112				192
10	62	95	83			240
20	21	73	102	61		257
30		40	63	73	25	201
40			40	52	18	110
Total	163	320	288	186	43	1000

(AEB)

Mathematical Note 7.1

Derivation of Spearman's Rank Correlation Coefficient   With the notation of page 84 the values of the X's and the Y's range from 1 to n, and the arithmetic mean of each is

$$\frac{n(n+1)}{2n} = \frac{n+1}{2} \qquad \left[\frac{n(n+1)}{2} \text{ is the sum of the A.P. } 1 \ldots n\right]$$

The deviation of the X rank from the mean is $X_k - \dfrac{n+1}{2} = x_k$ say, and

the corresponding individual in the Y rank $Y_k - \dfrac{n+1}{2} = y_k$ say. Therefore

$$X_k - Y_k = x_k - y_k$$

Now $r = \dfrac{S_{xy}}{S_x \times S_y} = \dfrac{\dfrac{\Sigma xy}{n} - \dfrac{\Sigma x}{n} \times \dfrac{\Sigma y}{n}}{\sqrt{\left(\left[\dfrac{\Sigma x^2}{n} - \left(\dfrac{\Sigma x}{n}\right)^2\right]\left[\dfrac{\Sigma y^2}{n} - \left(\dfrac{\Sigma y}{n}\right)^2\right]\right)}}$

If x and y are the deviations from the means, Σx and Σy equal zero.

Therefore $r = \dfrac{\dfrac{\Sigma xy}{n}}{\sqrt{\left(\dfrac{\Sigma x^2}{n} \times \dfrac{\Sigma y^2}{n}\right)}} = \dfrac{\Sigma xy}{\sqrt{(\Sigma x^2 \times \Sigma y^2)}}$

Let ρ stand for the rank correlation coefficient and write $\rho = \dfrac{\Sigma xy}{\sqrt{(\Sigma x^2 \times \Sigma y^2)}}$

where x and y are the rank deviations from the means of the ranks. Now

$$\Sigma x^2 = \Sigma\left(X - \frac{n+1}{2}\right)^2$$

$$= \Sigma X^2 - (n+1)\Sigma X + \frac{n(n+1)^2}{4}$$

ΣX^2 is the sum of the squares of the first n natural numbers, and

Therefore $\Sigma X^2 = 1^2 + 2^2 + \ldots n^2 = \tfrac{1}{6}n(n+1)(2n+1)$

$$\Sigma x^2 = \tfrac{1}{6}n(n+1)(2n+1) - (n+1)\frac{n(n+1)}{2} - \frac{n(n+1)^2}{4}$$

$$= \tfrac{1}{12}n(n+1)(n-1)$$

$$= \tfrac{1}{12}n(n^2-1)$$

Similarly $\Sigma y^2 = \tfrac{1}{12}n(n^2-1)$

If d_k is the difference between the ranks X_k and Y_k then

$$\Sigma d_k^2 = \Sigma(X_k - Y_k)^2$$
$$= \Sigma(x_k - y_k)^2$$
$$= \Sigma x^2 + \Sigma y^2 - 2\Sigma xy$$

(dropping the suffixes)

$$= \tfrac{1}{12}n(n^2-1) + \tfrac{1}{12}n(n^2-1) - 2\Sigma xy$$

Therefore $2\Sigma xy = \tfrac{1}{6}n(n^2-1) - \Sigma d^2$

Therefore
$$\rho = \frac{\Sigma xy}{\sqrt{(\Sigma x^2 \times \Sigma y^2)}}$$

$$= \frac{\dfrac{n(n^2-1)}{12} - \dfrac{\Sigma d^2}{2}}{\tfrac{1}{12}n(n^2-1)}$$

$$= 1 - \frac{6\Sigma d^2}{n(n^2-1)}$$

CHAPTER 8

Probability or Chance

The first part of this chapter is a résumé of definitions and theorems in *Elementary Statistics*.

Classical Definition

If an event can happen in a ways out of n equally possible occurrences, then the probability of the event happening is a/n. Similarly if an event can fail to happen in b ways out of n equally possible occurrences, then the probability of the event not happening is b/n. The letter p is used to denote a/n, and the letter q to denote b/n. Obviously $p+q = 1$.

Empirical or Relative Frequency Definition

If an event happens a times in n trials then

$$\underset{n \to \infty}{\mathrm{Lt}}(a/n) = p$$

This reads 'the limit of a/n as n tends towards infinity equals p'. 'n tends towards infinity' means a large number of trials should take place. What is 'large' must be determined by the nature of the problem.

Mutually Exclusive Events

Two events are said to be mutually exclusive when both cannot happen at the same time.

Theorems on Probability

Theorem 1. The theorem of total probability, sometimes known as the theorem of 'either . . . or'.

If two events are mutually exclusive, and the probability of the first event happening is p_1, and of the second event p_2, then the probability of either the first event or the second event happening is $p_1 + p_2$.

Theorem 2. The theorem of compound probability, sometimes known as the theorem of 'both . . . and'.

If the probability of one event happening is p_1, and of a second event p_2, then the probability of both the first and the second event happening is $p_1 \times p_2$.

Theorem 3. The 'at least one' theorem.

If the probabilities of n independent events are $p_1, p_2, p_3, \ldots p_n$, then the probability of at least one of the events happening is

$$1 - (1 - p_1)(1 - p_2) \ldots (1 - p_n)$$

Mathematical Expectation

Suppose p is the probability of a person winning a lottery prize of £S, then his expectation is defined as £pS. This definition can be extended. Thus the probability of obtaining 0, 1, 2, 3, or 4 heads if four coins are tossed is:

Heads	0	1	2	3	4
Probability	1/16	1/4	3/8	1/4	1/16

Therefore in 1600 trials the number of times we expect heads to appear 0, 1, 2, 3, 4 times is 100, 400, 600, 400 and 100 times, respectively. The total number of heads we expect is:

$$100 \times 0 + 400 \times 1 + 600 \times 2 + 400 \times 3 + 100 \times 4 = 3200$$

in the 1600 trials; and the average number expected per trial is 2. This is the expected value, or mean value, of the number of heads; i.e. in a large number of trials we expect the average value of heads to be near 2.

EXAMPLE

A, B, C, D cut a pack of cards in that order. The first to cut a spade wins a prize of £700. What are their expectations?

SOLUTION

For the first cut A's chance is $\frac{1}{4}$. If A fails, B's chance is $\frac{3}{4} \times \frac{1}{4}$, since A's chance of failing is $\frac{3}{4}$. Similarly the chance of C winning on the first cut is

$(\frac{3}{4})^2 \times \frac{1}{4}$, and D's chance $(\frac{3}{4})^3 \times \frac{1}{4}$. Similarly A's chance on a second cut is $(\frac{3}{4})^4 \times \frac{1}{4}$, and B's, C's and D's chances are $(\frac{3}{4})^5 \times \frac{1}{4}$, $(\frac{3}{4})^6 \times \frac{1}{4}$ and $(\frac{3}{4})^7 \times \frac{1}{4}$, respectively, and so on for further cuts. Thus

> A's chance is $\frac{1}{4}[1 + (\frac{3}{4})^4 + (\frac{3}{4})^8 + \ldots$ to infinity$] = 64/175$
> B's chance is $\frac{1}{4}[\frac{3}{4} + (\frac{3}{4})^5 + \ldots] = 48/175$
> C's chance is $\frac{1}{4}[(\frac{3}{4})^2 + (\frac{3}{4})^6 + \ldots] = 36/175$
> D's chance is $\frac{1}{4}[(\frac{3}{4})^3 + (\frac{3}{4})^7 + \ldots] = 27/175$.

Therefore
> A's expectation is $£700 \times 64/175 = £256$
> B's expectation is $£700 \times 48/175 = £192$
> C's expectation is $£700 \times 36/175 = £144$
> D's expectation is $£700 \times 27/175 = £108$
>
> ————
>
> Total $£700$

Thus if they decided not to compete but to share the prize money the above amounts are their expectations and would be a fair division of the prize money.

Exercise 8.1

1 Five cards are drawn at random one at a time from a pack and are not replaced. What is the probability that they are all spades?

2 Four cards are drawn from a pack and they are of different suits. What is the chance that they are drawn in (i) the order—club, diamond, heart, spade, (ii) in any order?

3 The chance of obtaining a defective ball-bearing is 5%. What is the chance of obtaining at least one in a sample of size (i) 5, (ii) 10?

4 The first, second and third prizes in a raffle are £100, £50 and £20, respectively, and the probability of winning a prize is 0·0001. What is a fair price to pay for a ticket?

5 A man has a 0·7 chance of winning £500 and a 0·3 chance of losing £100. If he wishes to sell his chance what should he expect to be paid?

An Alternative Notation

Notation is very important in mathematics and can simplify the development of a subject. At present the following notation seems to be gaining favour in probability.

If an event is named E then the probability of it occurring (what we call a success) is denoted by $P\{E\}$. The probability of its failure, or non-occurrence, is denoted by $P\{\text{not } E\}$ or $P\{\bar{E}\}$. Obviously $P\{E\} + P\{\bar{E}\} = 1$.

If E_1 and E_2 are two events, then the probability that E_2 occurs, given that E_1 has occurred, is written $P\{E_2|E_1\}$ and is called the conditional probability of E_2, given E_1 has occurred.

EXAMPLE

Suppose there are five white and three black balls in a box. Calculate the probability of drawing two black balls in succession if the balls are not replaced.

SOLUTION

Let E_1 denote the event of drawing the first black ball; then

$$P\{E_1\} = \tfrac{3}{8}$$

Let E_2 denote the event of drawing the second black ball; then

$$P\{E_2|E_1\} = \tfrac{2}{7}$$

Therefore the probability that both E_1 and E_2 occur is

$$\tfrac{3}{8} \times \tfrac{2}{7} = \tfrac{3}{28}$$

The probability that both E_1 and E_2 occur is written

$$P\{E_1E_2\}$$

and this equals

$$P\{E_1\}P\{E_2|E_1\}$$

If E_1 and E_2 are independent events, then

$$P\{E_1E_2\} = P\{E_1\}P\{E_2\}$$

This can be extended to three or more events. If E_1 and E_2 are mutually exclusive events, then

$$P\{E_1E_2\} = 0$$

for the occurrence of E_1 excludes the occurrence of E_2.

Sample Space

In modern theory each possible outcome of an experiment that can be repeated under similar conditions is called a sample point, and the set of sample points is said to occupy a sample space. Thus if two coins are tossed there are four sample points,

$$H_1H_2, H_1T_2, T_1H_2, T_1T_2$$

where H_1, T_1 represent heads and tails on the first coin and H_2, T_2 refer to the second coin. These four sample points make up the sample space. A set of points can be divided into subsets. Thus our set of four points may be divided into two subsets:

Subset A = those points containing at least one H
Subset B = those points containing at least one T

Obviously, two of the points belong to both subsets for

Subset $A = H_1H_2, H_1T_2, T_1H_2$
Subset $B = H_1T_2, T_1H_2, T_1T_2$

If $A + B$ is the combination of the sets denoted by A and B and AB is the set of points common to both A and B, then obviously

Combination of A and B = number of points in A + number of points in
B − number of points in AB,

since the number of points in AB is added twice. Therefore, Combination $(A + B) = A + B - AB$.

Euler or Venn Diagram

The subsets A and B can be represented in an Euler or Venn diagram.

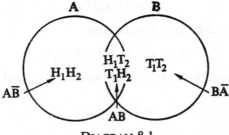

DIAGRAM 8.1

The subset A is represented by the left circle and the subset B by the right circle. The set of points common to both, (AB), is represented by the intersection of the circles. The points in A but not in B, $(A\bar{B})$, by the left lune, and the points in B but not in A, $(B\bar{A})$, by the right lune. If we divide the number of points in each section by the total number of points we get the probability of obtaining an event in that section. Thus,

$$P\{A\} = \tfrac{3}{4}; \quad P\{B\} = \tfrac{3}{4}; \quad P\{A\bar{B}\} = \tfrac{1}{4}; \quad P\{B\bar{A}\} = \tfrac{1}{4}; \quad P\{AB\} = \tfrac{1}{2}$$

It follows, $\qquad\qquad P\{A+B\} = P\{A\}+P\{B\}-P\{AB\}$ (8.1)

If A and B are mutually exclusive results then

$$P\{A+B\} = P\{A\}+P\{B\}$$

(Readers familiar with set notation will write the relation 8.1 in the form

$$P\{A \cup B\} = P\{A\}+P\{B\}-P\{A \cap B\}.)$$

The above can be extended to three or more subsets.

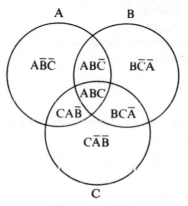

DIAGRAM 8.2

From a study of the diagram we get,

$$\begin{aligned}
A\bar{B}\bar{C} &= A-AB-CA+ABC \\
B\bar{C}\bar{A} &= B-AB-BC+ABC \\
C\bar{A}\bar{B} &= C-BC-CA+ABC \\
B\bar{C}\bar{A} &= BC-ABC \\
C\bar{A}\bar{B} &= CA-ABC \\
A\bar{B}\bar{C} &= AB-ABC \\
ABC &= ABC
\end{aligned}$$

The seven sections on the left of the equations give the total number of points in A, B, and C; so adding, we have

$$(A+B+C) = (A)+(B)+(C)-(AB)-(BC)-(CA)+(ABC)$$

or in terms of probability,

$$P\{A+B+C\} = P\{A\}+P\{B\}+P\{C\}-P\{AB\}-P\{BC\}-P\{CA\}+P\{ABC\}$$

EXAMPLE

Out of 257 students studying foreign languages it was known that the following numbers studied the different languages:

French	168	French and German	37
German	94	German and Spanish	22
Spanish	135	Spanish and French	98

Calculate the number of students studying

(a) all three subjects,
(b) French but not Spanish,
(c) German but not French,
(d) Spanish but not German,
(e) French or Spanish but not German,
(f) French but not German or Spanish.

SOLUTION

Let A, B, C denote the subsets of students studying French, German and Spanish, respectively. Then

$$(A+B+C) = (A)+(B)+(C)-(AB)-(BC)-(CA)+(ABC)$$

That is $257 = 168+94+135-37-22-98+(ABC)$

Therefore $(ABC) = 17$

It is now easy to construct a Venn diagram.

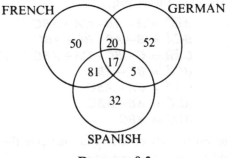

FRENCH   GERMAN

SPANISH

DIAGRAM 8.3

Thus 37 students study French and German, but 17 study all three languages. Therefore 20 must study French and German only. Repeating the reasoning we obtain the other figures in Diagram 8.3.

We can now determine the probability of any student picked at random studying any of the subjects or group of subjects. The probability of a student studying all three languages = 17/257. The probability of a student studying Spanish only = 32/257, and so on.

EXAMPLE

Set up a sample space for the toss of two tetrahedral dice. From the sample space determine the probability of obtaining a sum of (a) 6, (b) 3.

SOLUTION

The sample space is as follows. The first number is that on the face on which the first die rests; the second is that on which the second die rests.

$$
\begin{array}{cccc}
(1,1) & (2,1) & (3,1) & (4,1) \\
B\;(1,2) & (2,2) & (3,2) & (4,2)\;A \\
(1,3) & (2,3) & (3,3) & (4,3) \\
(1,4) & (2,4) & (3,4) & (4,4)
\end{array}
$$

There are 16 points in the sample space and the subset where the sum of the pairs of numbers is 6 is indicated by A, and $P\{A\} = 3/16 = 0\cdot1875$. B is the subset where the sum of the pairs of numbers is 3, and $P\{B\} = 2/16 = 0\cdot125$.

EXAMPLE

A black ball was drawn from one or other of two bags, one of which contained 3 black and 7 white balls, and the other 6 black and 5 white balls. Calculate the probability that the ball was drawn from the first bag.

SOLUTION

To help in the solution of this problem we will use the relative frequency definition of probability.

Suppose the experiment is repeated a great number of N times. If the bag from which the drawing is made is randomly chosen, the probability of choosing either bag is $\frac{1}{2}$. Therefore, in the long run each bag is chosen $\frac{1}{2}N$ times. In $\frac{1}{2}N$ drawings from the first bag there are $\frac{1}{2}N(3/10)$ drawings of a black ball, and from the second bag $\frac{1}{2}N(6/11)$ drawings of a black ball. Therefore in the N drawings there are $\frac{1}{2}N(3/10)+\frac{1}{2}N(6/11)$ drawings of a black ball, and $\frac{1}{2}N(3/10)$ of these are from the first bag. Therefore, the probability that the black ball is drawn from the first bag is

$$\frac{\dfrac{N}{2} \times \dfrac{3}{10}}{\left(\dfrac{N}{2} \times \dfrac{3}{10}\right) + \left(\dfrac{N}{2} \times \dfrac{6}{11}\right)}$$

$$= \frac{33}{33 + 60} = \frac{11}{31}$$

Let us express this problem in terms of the notation of page 95. Let E_1, E_2 denote the events of selecting the first or second bag, respectively, and A the event of selecting a black ball. Then $P\{E_1\}$ and $P\{E_2\}$ are the probabilities of selecting the first or second bag. In a large number of N trials the first bag is selected $NP\{E_1\}$ times, and the second bag $NP\{E_2\}$ times. The probability of selecting a black ball from the first bag is $P\{A|E_1\}$ and from the second bag $P\{A|E_2\}$. Therefore the number of times a black ball is selected from bag one is $NP\{E_1\} \times P\{A|E_1\}$ and from bag two $NP\{E_2\} \times P\{A|E_2\}$. Therefore the probability that knowing a black ball is selected, it came from bag one is

$$P\{E_1|A\} = \frac{P\{E_1\}P\{A|E_1\}}{P\{E_1\}P\{A|E_1\} + P\{E_2\}P\{A|E_2\}} \quad (N \text{ cancelling})$$

In our problem $P\{E_1\} = P\{E_2\} = \frac{1}{2}$; $P\{A|E_1\} = 3/10$; $P\{A|E_2\} = 6/11$.

Therefore $P\{E_1|A\} = \dfrac{\frac{1}{2} \times \frac{3}{10}}{\frac{1}{2} \times \frac{3}{10} + \frac{1}{2} \times \frac{6}{11}} = \frac{11}{31}$

and $P\{E_2|A\} = \dfrac{\frac{1}{2} \times \frac{6}{11}}{\frac{1}{2} \times \frac{6}{11} + \frac{1}{2} \times \frac{3}{10}} = \frac{20}{31}$

This result can be generalised. Suppose an observed event (A) has happened through one of mutually exclusive events E_1, E_2, ... E_n; then the probability of E_r being the cause of A is given by

$$P\{E_r|A\} = \frac{P\{E_r\}P\{A|E_r\}}{\sum\limits_{1}^{n} P\{E_r\}P\{A|E_r\}}$$

This is known as *Bayes' Theorem*.

Exercise 8.2

1   Set up a sample space for the toss of two coins and a six-sided die. From the sample space estimate the probability of obtaining

(a) two heads and a 1, 3 or 5 on the die,
(b) a head, a tail and a five on the die.

2 Three candidates *A*, *B* and *C* were nominated for three vacancies on a local council. Before the election the local newspaper held a sample poll to find its readers' opinions of the three candidates for the office of councillor. The following results were obtained. The numbers are percentages of the total number of replies received.

In favour of both A and B	16
In favour of A and B but not C	10
In favour of B but not A or C	20
In favour of A and C but not B	7
In favour of C but not B	38
In favour of B and C but not A	59

What percentage of readers were in favour of (a) A, (b) B, (c) C, (d) all three candidates, (e) any two out of the three candidates, (f) one only of the candidates?

3 Three players A, B and C throw two dice each, in the order ABCABCA ... until one of them throws a double six. Each player puts one pound into the pool at the beginning of the game, and the first player to throw a double six takes the pool. Find the expected net gain of the three players.

(AEB)

4 Four boys are taken at random. What is the probability that they were (i) all born on the same day of the week, (ii) all born on different days of the week? It may be assumed that births are equally likely on the seven days of the week.

If *n* boys are taken at random, what is the least value of *n* for which it is more likely than not that two at least were born on the same day of the week?

(AEB)

5 Three stamps are taken at random from a box containing 5 orange, 4 brown and 3 blue stamps. Find the probability that

(i) all three are of the same colour;
(ii) all three are of different colours;
(iii) two are of the same colour and the third of a different colour.

(L)

6   In a race the odds are 3 to 1 against A being in the first three, 8 to 1 against B being in the first three and 4 to 1 against C being in the first three. What are the odds for or against

(a) the first three places being filled in any order by A, B and C?
(b) at least one of A, B, C being in the first three?   (L)

7   One letter is selected from each of the names

JONES, THOMSON, WILKINSON

(a) Find the probability that the three letters are the same.
(b) Show that it is nearly fifteen times as probable that only two of the three letters are the same.   (L)

8   In a game for two players a turn consists of throwing a die either once or twice, once if the score obtained is less than 6, twice if the score at the first throw is 6. The score for the turn is in the first case the score of the single throw, and in the second case the total score of the two throws. Obtain the probabilities of a player

(a) scoring more than 9 in a single turn,
(b) scoring a total of more than 20 in two succeeding turns,
(c) obtaining equal scores in two succeeding turns.   (C)

9   In a television panel game each player is asked a series of questions until either he answers one incorrectly or he has answered three correctly; his score for the turn is the number of correct answers he has given, with a bonus mark if he gives three correct answers. Assume that the questions are independent and that the chance of giving a correct answer to any question is p. Obtain the mean score for a turn. Show that for $p = \frac{1}{2}$ the chance that a team of three players obtain a total score over 3 after one turn each is 25/64.   (C)

10   Two dice are thrown together and the scores added. What is the chance the total score exceeds 8? Find the mean and standard deviation of the total score. What is the standard deviation of the score for a single die?   (C)

11   A bag contains m oranges and n lemons; sampling is random without replacement. Obtain the probabilities of the following events:

(i) The first fruit drawn is an orange.
(ii) The first two fruits drawn are oranges.

(iii) The second fruit is a lemon, given that the first was an orange.

(iv) The third fruit is an orange.   (C)

12 Discuss carefully the meaning of the following statements:

(a) The probability that the next throw of a certain penny will show head is 0·5.

(b) The probability that at the next throw of the same penny the coin will stand on edge is 0.

(c) The probability that the height of the person in the next desk is 164 cm is 0·1.

(d) The probability that the height of the person in the next desk is 164 cm is 0.

(e) The probability that the next tall man I meet will have red hair is 0·05.

(f) The probability that Bacon wrote the plays of Shakespeare is 0·02.
                                                                      (C)

13 A farmer keeps two breeds A, B of chicken; 70% of the egg production is from birds of breed A. Of the eggs laid by the A hens, 30% are large, 50% standard and the remainder small: for the B hens the corresponding proportions are 40%, 30% and 30%. Egg colour (brown or white) is manifested independently of size in each breed; 30% of A eggs and 40% of B eggs are brown. Find:

 (i) the probability that an egg laid by an A hen is large and brown;

(ii) the probability that an egg is large and brown;

(iii) the probability that a brown egg is large;

(iv) which size grade contains the largest proportion of brown eggs;

 (v) whether colour and size are manifested independently in the total egg production.   (C)

14 An ordinary unbiased cubical die with faces marked 1 to 6 is thrown twice. Tabulate the probability distribution of the sum of two throws.

In a certain game a player makes three throws. At each throw he scores according to the face uppermost, if it is a six or if he has previously thrown a six, but not otherwise. Tabulate the probability distribution of his total score.

Show that his chance of scoring more than 11 is less than 1 in 6, but that his chance of scoring more than 10 exceeds 1 in 6.   (C)

15 Three dice are thrown. What is the probability of obtaining at least two sixes?

16   Three identical bags have two balls in each. The balls are identical except for colour. The first bag has two black balls, the second bag two white balls, and the third bag one black and one white ball. A ball is selected from a bag chosen at random and is found to be white. What is the probability the other ball in the bag is black?

17   In 10 games of golf between two players A and B, A won 5 games, B won 3 games and 2 games ended in a tie. What is the probability that in the next three games

(a) A wins the three games,
(b) B wins the three games,
(c) two of the games end in a tie,
(d) A and B win alternatively,
(e) B wins at least one game?

18   In a mixed school 15 boy and 10 girl prefects are to be selected from a sixth form of 60 boys and 50 girls. One of the boys is a brother of one of the girls. What is the probability

 (i) that both are selected,
(ii) that at least one is selected?   (AEB)

19(a)   A private telephone system uses ten independent units of apparatus to make each connection. The probability of failure of any one of the ten units is 0·01. Calculate the probability that any given call will be made at the first attempt.
    Find the probability that the call will not be correctly connected in five attempts.

(b)   If it rains one day, the probability that it rains the next day is 4/5, but if it does not rain, the probability that it rains the next day is 2/5. It does not rain on Tuesday. Calculate the probability that it will rain on Thursday of the same week.   (AEB)

20(a)   In a quiz programme, there are five competitors, each of whom makes a random choice of ten prizes, three of which are booby prizes. No two competitors can have the same prize. What is the probability (i) that none of the booby prizes is selected, (ii) that all three booby prizes are chosen?
(b)   Two poor marksmen, A and B, fight a duel. The probability that A will shoot B at any attempt is 1/10, and the probability that B will shoot

A at any attempt is 1/8. A shoots first, and then B and A shoot alternately until one or the other is hit, or until A and B have fired six shots each. What is the probability that neither A nor B is shot?   (AEB)

21   In an examination the respective probabilities of three candidates solving a certain problem are 4/5, 3/4, and 2/3. Calculate the probability that the examiner will receive from these candidates

  (i) one, and only one, correct solution,
 (ii) not more than one correct solution,
(iii) at least one correct solution.   (JMB)

22 (a)   An unbiassed die is tossed twice. What is the probability of obtaining

  (i) a six on either the first or second toss,
 (ii) at least one six?

(b)   A survey of 476 graduates studying one or more courses in Chemistry, Biology and Statistics gave the following number of students in the indicated subjects:

Chemistry	242	Chemistry and Biology	75
Biology	257	Biology and Statistics	161
Statistics	319	Chemistry and Statistics	157

What is the probability that a student selected at random takes all three subjects?   (AEB)

CHAPTER 9

Permutations and Combinations

Introduction

If the calculations of the probability of a complex event is laborious, the methods of permutations and combinations will often simplify the calculation. The proofs of the theorems quoted in this chapter are given in elementary mathematics books. n or $n!$ is called factorial n and is a short notation for the product of the factors $n, n-1, n-2, \ldots 3, 2, 1$. Thus $4! = 4 \times 3 \times 2 \times 1 = 24$. Remember $0! = 1$.

Permutations

When different arrangements of given items are made the items are said to be permuted. The number of permutations of n different things r at a time is written as $_nP_r$, or $P(n, r)$ or P_{nr}. We shall use $_nP_r$.

$$_nP_r = \frac{n!}{(n-r)!}$$

for example:
$$_5P_3 = \frac{5!}{2!} = 3 \times 4 \times 5 = 60$$

The number of permutations of n things taken all together, when there are p alike of one kind, q alike of a second kind, r alike of a third kind, and so on, is

$$\frac{n!}{p! \times q! \times r! \times . .}$$

EXAMPLE

How many permutations can be made from the letters of the word *statistics*?

106

SOLUTION

There are 10 letters, and s, t, a, i, c occur 3, 3, 1, 2, 1 times, respectively.

Therefore, the number of permutations $= \dfrac{10!}{3! \times 3! \times 2!} = 50\,400.$

The number of permutations of n things taken r at a time when each of the things may be repeated r times is n^r.

EXAMPLE

How many five-figure numbers can be formed from the digits 1 to 9 if each digit may be repeated any number of times?

SOLUTION

The first digit can be selected in 9 ways, and, since the digits can be repeated, the second digit can be selected in 9 ways, and so on for the rest of the digits. Therefore the number of numbers that can be formed is

$$9 \times 9 \times 9 \times 9 \times 9 = 9^5 = 59\,049$$

The total number of ways in which a selection may be made from n things which are not all different but which contain p alike of one kind, q alike of a second kind, and so on is

$$(p+1)(q+1) \ldots -1$$

omitting the case where all items are rejected.

EXAMPLE

How many selections can be made from 6 boys and 5 girls?

SOLUTION

We can select 1, 2, 3, 4, 5 or 6 boys or reject the lot. Therefore there are seven possibilities. Similarly there are six possibilities in the case of the girls.

Therefore the total number of possibilities is $7 \times 6 = 42$. This includes the case when all are rejected; therefore the required number is $42 - 1 = 41$.

Combinations

If r objects are selected out of n objects and no attention is paid to the order of selection, the selection is called a combination of r objects out of n. Thus a, b and b, a are two permutations but one combination. The number of combinations of n objects r at a time is denoted by $_nC_r$, or C_{nr}, or $\binom{n}{r}$ and is given by

$$_nC_r = \frac{n!}{(n-r)!\,r!}$$

EXAMPLE

Evaluate $_8C_5$.

SOLUTION

$$_8C_5 = \frac{8!}{3!\,5!} = 56$$

EXAMPLE

In how many ways can 24 different articles be divided into groups of 12, 8 and 4 articles, respectively?

SOLUTION

The first group can be selected in $_{24}C_{12}$ ways.
The second group can then be selected in $_{12}C_8$ ways.
The remaining four articles form the third group.
Therefore the number of ways the articles can be divided is

$$_{24}C_{12} \times {}_{12}C_8 = \frac{24!}{12!\,12!} \times \frac{12!}{8!\,4!} = 1\ 338\ 557\ 220$$

EXAMPLE

What is the probability that, in a card game involving four players, a dealer of a pack of 52 playing cards receives the whole of one suit?

SOLUTION

The number of ways thirteen cards may be selected from fifty-two is $_{52}C_{13}$. In these $_{52}C_{13}$ ways there will be once in each case 13 clubs, 13 diamonds, 13 hearts and 13 spades. Therefore the chances of 13 of one suit appearing is $4/_{52}C_{13}$. The chances of the dealer receiving this suit is $\frac{1}{4}$. Therefore the probability he deals himself a complete suit is

$$\frac{1}{4} \times \frac{4}{_{52}C_{13}} = 1 \cdot 212 \times 10^{-13}$$

EXAMPLE

There are three black balls in a bag of fifty balls. The balls are drawn in succession (without replacement) from the bag. Calculate the chance that the first black ball is the rth ball.

SOLUTION

The chance the first ball drawn is not black $= 47/50$
The chance the second ball drawn is not black $= 46/49$

.
.
.

The chance the $(r-1)$th ball drawn is not black $= \dfrac{47-(r-2)}{50-(r-2)}$

The chance the rth ball drawn is black $= \dfrac{3}{50-(r-1)}$

Therefore the probability the rth ball drawn is the first black ball

$$= \frac{47}{50} \times \frac{46}{49} \times \frac{45}{48} \cdots \frac{[47-(r-2)]}{[50-(r-2)]} \times \frac{3}{[50-(r-1)]}$$

Now $\qquad 47 \times 46 \ldots [47-(r-2)] = 47!/(48-r)!$
and $\qquad 50 \times 49 \ldots [50-(r-2)] = 50!/(51-r)!$

Therefore, the probability is

$$\frac{47!}{(48-r)!} \times \frac{(51-r)!}{50!} \times \frac{3}{51-r}$$

$$= \frac{3(49-r)(50-r)(51-r)}{48 \times 49 \times 50 \times (51-r)} \quad \text{(cancelling)}$$

$$= \frac{(49-r)(50-r)}{39\,200}$$

Exercise. 9.1

1   How many different arrangements are there of all the letters of the word SYNONYM?

Counting Y as a vowel, how many arrangements have consonants and vowels alternately?

If an arrangement is taken at random, what is the probability that the third letter is an N?

If we are told that the arrangement has consonants and vowels alternately and the first letter is an S, what is the probability that the third letter is an N?   (AEB)

2   Evaluate (i) $_6P_3$. (ii) $_8C_6$.

Calculate the value of n if $_nP_2 = 20$.

From a mixed form of 13 girls and 17 boys, 2 girls and 3 boys have to be selected. In how many ways can this be done?   (AEB)

3   Find the number of different sequences of $(p+q)$ letters, p of them being E and q of them being N.

Six roads in a Garden City run east and west and are intersected by six avenues running north and south. Find the number of different routes, each of the same minimum length, which a man may take in walking from the extreme S.W. intersection to the extreme N.E. one.

Find, also, the chance that the man will pass a building situated on the third road (counting from the south) mid-way between its intersections with the third and fourth avenues (counting from the west).   (L)

4   A crew of eight oarsmen is arranged so that the crew row alternately right and left with the first man on the right.

(a) If half the crew can row only on the right, find the number of possible arrangements.

(b) If a crew of eight is to be chosen from 12 men of whom 4 can row only on the right, 5 can row only on the left and 3 can row on either side, find the number of different crews that can be chosen to include any 3 out of the 4 who can row only on the right. (A change in the order of the oarsmen is to be regarded as a change in the crew.)   (L)

5   Each of two bags contains 8 coins. How many different combinations of 6 coins can be made by drawing 3 coins out of each bag, (a) when

no two of the 16 coins are alike, (b) when two of the coins in one bag are alike?

In case (a) what is the probability that a specified coin will appear in any one combination? (L)

6   A box of chalks contains 5 white, 4 green, 2 red, 1 yellow. In how many different ways can they be arranged side by side?

Calculate also the probability that the 4 green chalks are together. (Consider chalks of the same colour as identical; leave your answer in factorials.) (L)

7   How many groups of 5 letters, all different, may be chosen from an alphabet of 26 letters? In how many of these groups will one specified letter occur?

Five different letters are written down at random. What is the probability that they are consecutive letters of the alphabet in their correct order? (L)

8   Find the number of different arrangements of all the letters of the word 'esteem' taken altogether.

If three letters are chosen at random from the word 'esteem', find the probability that at least two of them are e's. (L)

9   An encyclopaedia consisting of n ($\geqslant 4$) similar volumes is kept on a shelf with the volumes in correct numerical order; that is, with volume one on the left, volume two next and so on. The volumes are all taken down for cleaning and replaced on the shelf in random order. Prove that the probabilities of finding exactly n, $(n-1)$, $(n-2)$, $(n-3)$ volumes in their correct positions on the shelf are

$$\frac{1}{n!}, \quad 0, \quad \frac{1}{2(n-2)!}, \quad \frac{1}{3(n-3)!}$$

respectively. (C)

10(i)   Three cards are drawn at random from the 52 cards of a pack which contains 4 Kings, 4 Queens and 4 Knaves. Find the chance that one will be a King, one a Queen and one a Knave.

(ii)   A team of 6 qualifies for an athletic event if 4 of its members can succeed in a particular trial. Each member is allowed to attempt this trial three times. If the chance of success by any member of the team in any one attempt is 2/5, find the chance of the team qualifying. (O and C)

11(i)   From a bag containing five red, four white and three green balls three are drawn together at random. Find the chance of their being

(a) all of different colours,
(b) all of the same colour.

(ii)   Two six-faced dice whose faces bear the numbers 1 to 6, respectively are thrown together. Find the chance of

(a) the total score being exactly 8,
(b) the total score being greater than 8.   (O and C)

12(i)   A pack of 52 cards contains 4 Aces and 4 Kings. Three cards are taken at random from the pack. Find the chance that they should be (a) 3 Aces, (b) 2 Aces and 1 King.

(ii)   The chance of any one engine of a four-engined aeroplane failing on a long journey is 5 per cent. If an engine fails the chance of the aeroplane completing the journey is 80 per cent. It cannot fly with two engines out of action on the same wing. Find the chance that the aeroplane will complete the journey.

(The probabilities given in this question bear no relation to those that may exist on regular air services.)   (O and C)

13   A batch of 20 articles contains 3 which are substandard. Articles are drawn at random from the batch and are not replaced. Show that the chance that the first substandard article will be found at the rth draw is

$$P(r) = (20-r)(19-r)/2280$$

Find the value of $P(r)$ if each article is replaced before the next draw.
                                                                              (O and C)

14   Prove that the number of permutations of n unlike things r at a time, is $n!/(n-r)!$.

Nine plots of land are to be used in an agricultural experiment and five different manurial treatments are to be given to five of the plots, the remaining four plots being left untreated. Calculate

  (i) the number of ways in which the manurial treatments can be arranged,
 (ii) the probability that a given plot will receive some sort of manurial treatment,
(iii) the probability that each of two given plots will receive specified manurial treatments.   (JMB)

CHAPTER 10

The Binomial Distribution

General Deduction of the Binomial Distribution

In Chapter 20 of *Elementary Statistics* it was shown how to derive the binomial distribution in particular cases, and the mean np, and standard deviation $\sqrt{(npq)}$ for the general distribution were deduced. The coefficients of the terms of the distribution were obtained by means of Pascal's Triangle, but it is quite easy by means of the theorems on probability and combinations to obtain the general distribution. The important proviso in the distribution is that the chance of success of an event remains constant throughout the trials. If the chance of success is p and the chance of failure q, where $p+q = 1$, the probability of getting r successes and $n-r$ failures in n trials is by the theorem on Compound Probability $p^r q^{n-r}$. The number of ways of selecting r events out of n possible events is $_nC_r$. Hence the total probability of r successes and $n-r$ failures is $_nC_r p^r q^{n-r}$. If the probability of r successes is denoted by $P(r)$, we have

$$P(r) = {_nC_r}p^r q^{n-r}$$

We can have any number of successes from 0 to n, and therefore r can take any of the values from 0 to n. Therefore

$$(q+p)^n = P(0)+P(1)+P(2)+ \ldots +P(r)+ \ldots P(n)$$
$$= {_nC_0}p^0 q^n + {_nC_1}p^1 q^{n-1} + {_nC_2}p^2 q^{n-2} + \ldots + {_nC_n}p^n q^0$$

Now $p^0 = 1 = q^0$, and therefore

$$(q+p)^n = {_nC_0}q^n + {_nC_1}p^1 q^{n-1} + \ldots + {_nC_r}p^r q^{n-r} + \ldots + {_nC_n}p^n$$

This is the Binomial Expansion, and since

$$_nC_r = \frac{n!}{(n-r)!\,r!} \quad \text{and} \quad 0! = 1$$

113

the expression can be written

$$(q+p)^n = q^n + nq^{n-1}p + \frac{n(n-1)}{1\times 2}q^{n-2}p^2 + \ldots$$

$$+ \frac{n(n-1)(n-2)\ldots(n-r+1)}{1\times 2\times 3\times\ldots r}q^{n-r}p^r + \ldots + p^n$$

For mathematical reasons it is sometimes more convenient to write the expansion in the following manner,

$$(q+pt)^n = q^n + nq^{n-1}pt + \frac{n(n-1)}{1\times 2}q^{n-2}p^2t^2 + \ldots + p^nt^n$$

and to say that the probability of exactly r successes is the coefficient of t^r in the expansion. Thus the probability of exactly two successes is the coefficient of t^2 which is $\dfrac{n(n-1)}{1\times 2}q^{n-2}p^2$.

This expansion reduces to the original expansion if $t = 1$.

$(q+pt)^n$ is called the probability generating function.

Since $q = 1-p$, the generating function can be written as $(1-p+pt)^n$.

Calculation of the Terms of the Binomial Probability Function

The expansion of the binomial function can be calculated quite easily by means of Pascal's Triangle if n is small, but if n is large it is better to proceed as follows:

$$\frac{P(r+1)}{P(r)} = \frac{{}_nC_{r+1}q^{n-(r+1)}p^{r+1}}{{}_nC_rq^{n-r}p^r} = \frac{n-r}{r+1}\times\frac{p}{q}$$

Therefore, $P(r+1) = \dfrac{n-r}{r+1}\times\dfrac{p}{q}\times P(r)$

and taking logarithms,

$$\log P(r+1) = \log(n-r) - \log(r+1) + \log(p/q) + \log P(r) \qquad (10.1)$$

Suppose we wish to expand $(q+p)^{20}$, where $q = 0.9$ and $p = 0.1$.

$$P(0) = q^{20} = (0.9)^{20} = 0.1216 \quad \text{(using 7-figure logs)}$$
$$\log(p/q) = \log(1/9) = \bar{1}.0457$$

To calculate $P(1)$ we must put $r = 0$ in Equation (10.1), and

$$\log P(1) = \log(20-0)-\log(0+1)\log(1/9)+\log P(0)$$
$$= 1\cdot3010-0+\bar{1}\cdot0457+\bar{1}\cdot0848$$
$$= \bar{1}\cdot4315$$

Therefore $P(1) = 0\cdot2701$. To calculate $P(2)$ we put $r = 1$ in 10.1, and

$$\log P(2) = \log(20-1)-\log(1+1)+\log(1/9)+\log P(1)$$
$$= 1\cdot2788-0\cdot3010+\bar{1}\cdot0457+\bar{1}\cdot4315$$
$$= \bar{1}\cdot4550$$
$$P(2) = 0\cdot2851$$

and so on.
The work is best arranged in the form of a table.

Probability	r	$\log(n-r)$	$\log(r+1)$	$\log(p/q)$	$\log P(r+1)$	Probability
$P(0)$					$\bar{1}\cdot0848$	0·1216
$P(1)$	0	1·3010	0	$\bar{1}\cdot0457$	$\bar{1}\cdot4315$	0·2701
$P(2)$	1	1·2788	0·3010	$\bar{1}\cdot0457$	$\bar{1}\cdot4550$	0·2851
$P(3)$	2	1·2553	0·4771	$\bar{1}\cdot0457$	$\bar{1}\cdot2789$	0·1901
$P(4)$	3	1·2304	0·6021	$\bar{1}\cdot0457$	$\bar{2}\cdot9529$	0·0896
$P(5)$	4	1·2041	0·6990	$\bar{1}\cdot0457$	$\bar{2}\cdot5037$	0·0318
$P(6)$	5	1·1761	0·7782	$\bar{1}\cdot0457$	$\bar{3}\cdot9473$	0·0088
$P(7)$	6	1·1461	0·8451	$\bar{1}\cdot0457$	$\bar{3}\cdot2940$	0·0020
$P(8)$	7	1·1139	0·9031	$\bar{1}\cdot0457$	$\bar{4}\cdot5505$	0·0004
$P(9)$	8	1·0792	0·9542	$\bar{1}\cdot0457$	$\bar{4}\cdot7212$	0 0001
					Total	0·9996

Thus, to obtain $\log P(7)$, say, add 1·1461, subtract 0·8451, add $\bar{1}\cdot0457$, and add $\bar{3}\cdot9473$.

There should be twenty-one terms in the expansion, but with four-figure logarithms it is impossible to proceed further. The chances of obtaining nine successes works out to approximately one in ten-thousand, which is quite small. The sum of the probabilities should be one, but owing to the fourth figure of the probabilities being approximate, and the neglect of the probabilities of the last twelve terms, the total probability is 0·9996. Since $P(r)$ is the probability of obtaining r successful events in a trial consisting of n events, it follows that in k trials the number of trials with r successful events is $k \times P(r)$.

Thus if in the last example the trial was repeated 10 000 times the theoretical frequency of the number of successful events would be:

Successes	0	1	2	3	4	5	6	7	8	9	10
Frequency	1216	2701	2851	1901	896	318	88	20	4	1	0

If the arithmetic mean and the standard deviation of the above distribution are calculated, they will be found to be 2 and 1·342, respectively, agreeing with the theoretical values np and $\sqrt{(npq)}$.

Variations in the Value of p

The binomial distribution $(q+p)^n$ can vary with

(i) the value of p, and
(ii) the value of n.

Table 10.1 gives the frequencies of the number of successes in the expansion of $1000(q+p)^{20}$ when p varies from 0·1 to 0·9.

Table 10.1

Number of successes	$p =$ $q =$	0·1 0·9	0·2 0·8	0·3 0·7	0·4 0·6	0·5 0·5	0·6 0·4	0·7 0·3	0·8 0·2	0·9 0·1
0		122	12	1						
1		270	58	7	1					
2		285	137	28	3					
3		190	205	72	12	1				
4		90	218	130	35	5				
5		32	175	179	75	15	1			
6		9	109	192	124	37	5			
7		2	55	164	166	74	15	1		
8			22	114	180	120	36	4		
9			7	65	160	160	71	12		
10			2	31	117	176	117	31	2	
11				12	71	160	160	65	7	
12				4	36	120	180	114	22	
13				1	15	74	166	164	55	2
14					5	37	124	192	109	9
15					1	15	75	179	175	32
16						5	35	130	218	90
17						1	12	72	205	190
18							3	28	137	285
19							1	7	58	270
20								1	12	122
Total		1000	1000	1000	1001	1000	1001	1000	1000	1000

The sum of each series should be 1000, but there is, in some cases, a small discrepancy due to rounding off the numbers to the nearest integer.

Diagram 10.1 gives the frequency polygons for the binomial series $1000(q+p)^{20}$ for values of p from 0·1 to 0·9.

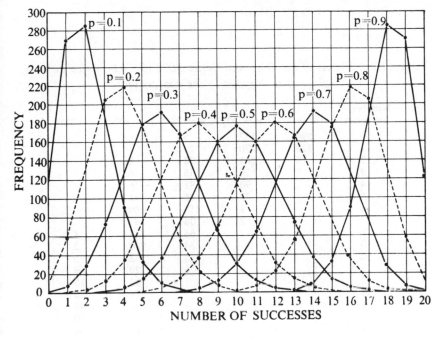

DIAGRAM 10.1

It will be noticed that for $p = 0.5$ the series is perfectly symmetrical and that for values of p less than 0.5 the series is positively skew. The series for p greater than 0.5 are the reflections of the series for p less than 0.5 and they are negatively skew. The arithmetic mean, the median and the mode increase as the value of p increases.

Variations in the Value of n

If we keep p constant and vary n we obtain values of the frequencies as shown in Table 10.2 and Diagram 10.2.

It will be noticed that as n increases the values of the arithmetic mean and the standard deviation increase and the series becomes more symmetrical.

It is important to remember that if (i) p approaches 0.5, or (ii) n becomes large, then the series tends to become symmetrical.

Table 10.2   *Terms of the Binomial Series 1000 $(0.9 + 0.1)^n$ for different values of* n. *Figures to the nearest unit*

Number of successes	$n = 5$	$n = 20$	$n = 50$	$n = 100$	$n = 200$
0	590	122	5		
1	328	270	29		
2	73	285	78	2	
3	8	190	138	6	
4		90	181	16	
5		32	185	34	
6		9	154	60	
7		2	108	89	
8			66	115	1
9			34	130	2
10			15	132	5
11			6	120	9
12				99	15
13				74	25
14				51	36
15				33	50
16				19	64
17				11	77
18				5	88
19				3	93
20				1	94
21					89
22					81
23					69
24					57
25					44
26					33
27					24
28					16
29					11
30					7
31					4
32					2
33					1
34					1
Total	999	1000	999	1000	998

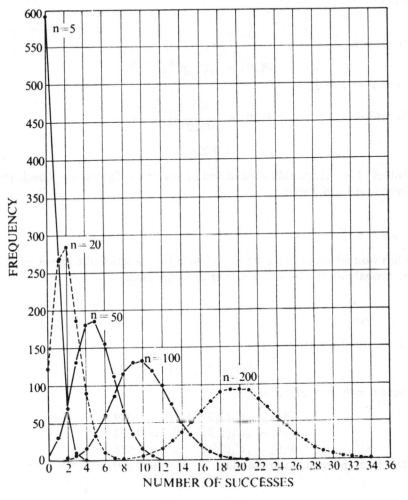

DIAGRAM 10.2

EXAMPLE

Thirty coins are tossed. What is the probability of getting exactly twenty heads?

SOLUTION

$$P(20) = {}_{30}C_{20}p^{20}q^{30-20}$$

where $p = q = \frac{1}{2}$.

Therefore
$$P(20) = \frac{30!}{10!\,20!}(\tfrac{1}{2})^{30}$$

$$= 0\cdot028$$

(Tables of log $n!$ are published in sets of Statistical Tables, e.g. *Lanchester Short Statistical Tables*.)

EXAMPLE

A has won 20 out of 30 games of golf with B. In a series of 6 games what is the probability that A will win four or more games?

SOLUTION

The required probability is

$$P(4) + P(5) + P(6)$$

where $n = 6$; $p = 2/3$; $q = 1/3$.

$$P(4) = {}_{6}C_{4}(2/3)^{4}(1/3)^{2} = 0\cdot329$$
$$P(5) = {}_{6}C_{5}(2/3)^{5}(1/3)\ = 0\cdot263$$
$$P(6) = {}_{6}C_{6}(2/3)^{6}\qquad = 0\cdot088$$

Therefore, the required probability $= 0\cdot680$

EXAMPLE

The probability of hitting a target with one bomb is $\frac{1}{2}$. It is required to have at least a 90% chance of destroying the target. If three direct hits are required to destroy the target completely, how many bombs must be dropped?

SOLUTION

Suppose n bombs are dropped. The probability of hitting the target with three or more bombs is

$$P(3) + P(4) + P(5) + \ldots + P(n)$$

and this must equal (90/100) or 0·9.

Now $P(0) + P(1) + P(2) + \ldots + P(n) = 1$

Therefore, $P(3) + P(4) + \ldots + P(n) = 1 - P(0) + P(1) + P(2)$

$$= 1 - {}_nC_0q^n + {}_nC_1q^{n-1}p + {}_nC_2q^{n-2}p^2$$
$$= 1 - (\tfrac{1}{2})^n[1 + n + \tfrac{1}{2}n(n-1)]$$
$$= 1 - (\tfrac{1}{2})^n[\tfrac{1}{2}(2 + n + n^2)]$$
$$= 0·9$$

Therefore, $\qquad (2 + n + n^2)/2^{n+1} = 0·1$

or, $\qquad\qquad\qquad 2^{n+1} = 10(2 + n + n^2)$

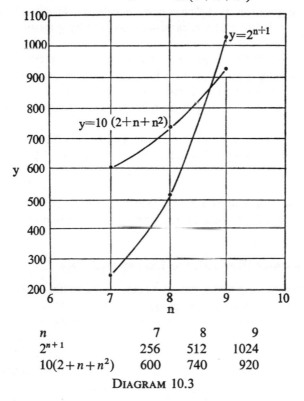

n	7	8	9
2^{n+1}	256	512	1024
$10(2 + n + n^2)$	600	740	920

DIAGRAM 10.3

To solve this equation plot $y = 2^{n+1}$ and $y = 10(2+n+n^2)$ on the same graph. Where the curves intersect gives the value of n.

The curves intersect between where $n = 8$ and $n = 9$. Therefore 9 bombs are required.

Exercise 10.1

1   A manufacturer has to reject 5% of the articles he manufactures. What are the probabilities of 0, 1, 2, 3, 4 rejects in a random sample of 10 articles?

2   The chances are that out of 100 babies born 51 will be boys. If 6 babies are born on the same day in a hospital, what is the probability of five of them being boys?

3   In a test of ten questions the candidate is required to answer 'Yes' or 'No' to each question. If the pass mark is four out of ten correct answers, what is the probability of the candidate passing by guessing?

4   To win a football competition it is necessary correctly to forecast 14 results. Assuming the chances of a win, draw or loss are equally probable, how many coupons must be filled in to make sure of one correct result?
    What are the chances of getting a coupon completely wrong?

5   A gambler tosses a number of coins and to win he must toss three or more heads. How many coins must he toss to have a 60% or better chance of winning?

6   How many cards must a man select from a pack of 52 playing cards to have at least a 70% chance of drawing three or more clubs?

Fitting a Binomial Distribution to Experimental Data

At times we may wish to determine whether a theoretical binomial distribution will represent experimental data. To do this we calculate the theoretical frequencies using the constants n and p obtained from the experimental results. If the experimental results are expected to fit a binomial distribution and fail to do so, it may be due to experimental errors such as the selection of items not being random. In this case the experimental technique should be checked to find the cause of the disagreement.

EXAMPLE

A gardener purchases some bulbs and he plants five in each of 100 bowls. After six weeks he obtains the following results:

Number of bulbs germinating	0	1	2	3	4	5
Number of bowls	1	6	14	33	31	15

Calculate the probability of a bulb germinating in six weeks and the theoretical frequencies if a binomial distribution is used to represent the results.

SOLUTION

The mean number of bulbs germinating is

$$(1 \times 0 + 6 \times 1 + 14 \times 2 + 33 \times 3 + 31 \times 4 + 15 \times 5)/100$$
$$= 3 \cdot 32$$

$n = 5$ and mean $= np$. Therefore $p = 3 \cdot 32/5 = 0 \cdot 664$, and $q = 0 \cdot 336$. The theoretical frequencies are given by the terms of $100(q+p)^5$, i.e.

$$100(q^5 + 5q^4 p + 10q^3 p^2 + 10q^2 p^3 + 5qp^4 + p^5)$$
$$= 0 + 4 + 17 + 33 + 33 + 13 = 100$$

The standard deviation $= \sqrt{(npq)} = \sqrt{(5 \times 0 \cdot 664 \times 0 \cdot 336)} = 1 \cdot 06$. At present we can only guess if the theoretical distribution is a good fit with the experimental data, but later we will describe a test—the χ^2 test—to measure the goodness of fit.

Exercise 10.2

1 The number of male live births in families of five live births are given in the table. Fit a theoretical binomial distribution to the results.

Number of Male Live Births in Families of Five
Live Births (Fictitious figures)

Number of Male Live Births	Number of Families
0	124
1	621
2	1301
3	1375
4	741
5	162
Total	4324

2   Two varieties of sweet peas are distinguishable by the colour of their seeds, one being white and the other green. From the next generation of seed pods 200 pods each containing four seeds are collected. The number of green seeds in the pods are given in the following table:

Number of green seeds	0	1	2	3	4	
Number of pods	10	56	78	44	12	= 200

Calculate the probability of obtaining a green seed and the terms of a theoretical binomial distribution with the same mean and total frequency.

3   Groups of nine flies are placed in one thousand containers and sprayed with an insecticide. After a given time the number of dead flies in each container is counted and the results are given in the table.

Number of dead flies	0	1	2	3	4	5	6	7	8	9	
Number of containers	147	287	259	180	79	33	8	3	2	2	= 1000

Calculate the probability of a fly being killed by the insecticide in the given time and fit a theoretical binomial distribution.

4   The number of males in 100 litters of 8 pigs are given in the table.

Number of males	0	1	2	3	4	5	6	7	8	
Number of litters	0	3	9	22	24	25	13	3	1	= 100

Fit a theoretical binomial distribution.

5   Write down the value of the arithmetic mean and standard deviation of a binomial distribution in which $n = 16$ and $p = \frac{1}{2}$.
  The probability of a pupil in a school not playing games is 1/5. The school is large and has 60 forms. If a random sample of 6 pupils is taken from each form in how many samples would it be expected to find (i) no players, (ii) exactly 3 players? (AEB)

6(i)   If 2 cards are drawn at random from a complete pack, what is the probability that they are both aces?

(ii)   If win, loss or draw are regarded as equally probable results of any Test Match, what is the probability that one side will win exactly 3 matches in a series of 5? (L)

7(i)   Three pennies are tossed simultaneously. Write down the probability that they will all come down 'heads', and the probability that there will be two 'heads' and one 'tail'.

(ii)   If a boy with four pennies and a girl with three pennies toss them at the same time, find the probability

(a) that each will throw only one 'head',
(b) that the boy will throw the same number of 'heads' as the girl.   (L)

8(a)   In a binomial distribution where the frequencies of occurrence of an event are given in terms of the expansion of $(q+p)^m$, the mean μ and the standard deviation σ of the distribution are given by the formulae

$$\mu = mp, \qquad \sigma = \sqrt{(mpq)}$$

Prove these formulae in the case where $m = 3$.

(b)   Two men A and B play a game in which A should win 8 games to every 7 won by B. If they play 3 games, show that the probability that A will win at least two games is approximately 0·55.   (C)

9   A battery of four guns is firing on to a target. It is reckoned that each gun should score on the average one direct hit in every five shots, and that three direct hits are needed to destroy the target. If each gun fires one shell, calculate the probability the target will be destroyed.

   With new gun crews it is reckoned that two of the guns should score one direct hit in every three shots and that the other two guns should score one direct hit every four shots. If each gun now fires one shell calculate the probability that the target will be destroyed.   (C)

10   In a certain competition for teams of six the qualifying event consists of each member of the team being allowed up to three attempts at a particular trial, the team as a whole qualifying if at least four members out of the six succeed in the trial. Assuming that the chance of failure in any one attempt by any member of the team is q, evaluate the chance of the team qualifying.   (C)

11   A 'one-armed bandit' advertised to 'Increase your money tenfold', costs sixpence a turn; the player is returned 5 shillings if more than 8 balls out of a total of 10 drop in a specified slot. The chance of any one ball dropping is p. Determine the chance of winning in a given turn, and for $p = 0.65$ calculate the mean profit made by the machine on 500 turns.

   Evaluate the proportion of losing turns in which the player comes within one or two balls of winning ($p = 0.65$).   (C)

12   Write down the binomial distribution for the number of successes in n independent trials when the chance of success in a single trial is p. Prove that the mean of the distribution is np.

(i) In a trial 8 coins are tossed together. In 100 such trials how many times should one expect to obtain 3 'heads' and 5 'tails'?

(ii) If 8 per cent of articles in a large consignment are defective, what is the chance that a sample of 30 articles will contain fewer than 3 defectives?

(O and C)

13   Show that the probability of r successes in n independent trials is the coefficient of t^r in the expansion of $(q+pt)^n$, where p is the chance of success in a single trial and $q = 1-p$. Prove that the mean number of successes is np.

Samples, each of 8 articles, are taken at random from a large consignment in which 20 per cent of articles are defective. Find the most likely number of defective articles in a single sample and the chance of obtaining precisely this number.

If 100 samples of 8 are to be examined, calculate the number of samples in which you would expect to find three or more defective articles.

(O and C)

14   Show that if a random sample of n articles is drawn from a large number of articles of which a proportion p is defective, the chance of the sample containing r defective articles is

$$_nC_rp^r(1-p)^{n-r}$$

A large batch of manufactured articles is accepted if either of the following conditions is satisfied:

(a) A random sample of 10 articles contains no defective articles.

(b) A random sample of 10 contains one defective article and a second random sample of 10 is then drawn which contains no defective articles.

Otherwise the batch is rejected.

If, in fact, 5 per cent of articles in a batch to be examined are defective find the chance of the batch being accepted. $[(0 \cdot 95)^{10} = 0 \cdot 5987.]$

(O and C)

15   Four out of a batch of 40 manufactured articles are known to be defective. If a sample of four is drawn at random from the batch, find the chance that it will contain (i) exactly two defective articles, (ii) not more than one defective article.

Find also the chance that, if the articles are examined successively, the third article should be the first to be found defective.   (O and C)

16 If b is numerically less than a, and m, n are integers, show that the first two terms in the binomial expansion of $(a+b)^{m/n}$ are

$$a^{(m/n)} \quad \text{and} \quad (m/n)a^{(m/n)-1}b$$

Write down the next two terms.
Hence evaluate $(32\cdot01)^{2/5}$ correct to nine places of decimals.   (AEB)

17 Samples of forty articles are selected at random from a large bulk of articles produced by a machine. The following list shows the number of defective articles in each of twenty such samples:

$$1, 0, 4, 0, 0, 4, 4, 1, 4, 2, 3, 3, 2, 3, 0, 2, 4, 2, 0, 1$$

Calculate the mean number of defective articles per sample and estimate the percentage of defective articles in the total output of the machine.
Assuming the binomial law applies,

(i) calculate the probability of there being one or more defective articles in a sample of forty,
(ii) find the least integer N such that the probability of there being N or more defective articles in a sample of forty is less than $\frac{1}{2}$.

Supposing that 10% of the whole bulk is sampled in the above way and that the defective articles found in each sample are rejected, estimate the percentage of defective articles finally remaining in the bulk.   (JMB)

18 A bag contains 12 white balls and 8 black balls. Calculate the probability that a random sample of 5 balls drawn together from the bag will contain at least 4 white balls.   (JMB)

19 If the probability of an event occurring at a single trial is p, prove that the probability of exactly r occurrences of the event in n independent trials is

$$_nC_rp^r(1-p)^{n-r}$$

Present records show that 10 per cent of 14-year-old children have had their tonsils removed. If a large number of 14-year-old children are examined in groups of thirty and a frequency distribution is drawn up showing the number in each group who have had their tonsils removed, estimate the mean, the variance and the mode of the distribution. (JMB)

Mathematical Note 10.1

The Arithmetic Mean and Standard Deviation of the Binomial Distribution.
The shortest method of proving that the arithmetic mean and standard
deviation of a binomial distribution are np and $\sqrt{(npq)}$, respectively, is by
using the probability generating function $(q+pt)^n$.

The probability of r successes is the coefficient of t^r in the expansion of
$(q+pt)^n$, i.e. $_nC_rq^{n-r}p^r$. If the total number of trials is k, the frequency of r
successes is $k_nC_rq^{n-r}p^r$. That is if f_r is the frequency of r successes,

$$f_r = k_nC_rq^{n-r}p^r$$

$k(q+pt)^n$ can now be expanded as

$$k(q+pt)^n = f_0t^0 + f_1t^1 + \ldots f_rt^r + \ldots + f_nt^n$$

Putting $t = 1$, we have

$$k = \sum_0^n f_r$$

i.e. the sum of the frequencies of the number of successes is k, which is
what we expect.

Differentiating both sides of Equation (10.2) with respect to t, we have

$$knp(q+pt)^{n-1} = 0 \times f_0 + 1 \times f_1t^0 + 2 \times f_2t^1r \ldots n \times f_nt^{n-1} \quad (10.3)$$

Putting $t = 1$ and since $(q+p)^{n-1} = 1$, we have

$$knp = 0 \times f_0 + 1 \times f_1 + 2 \times f_2 + \ldots n \times f_n$$
$$= \sum_0^n rf_r$$

Therefore $$np = \frac{\Sigma rf_r}{k}$$

$$= \text{mean number of successes}$$

Multiply Equation (10.3) by t

$$knpt(q+pt)^{n-1} = 0 + 1 \times f_1t + 2 \times f_2t^2 + \ldots n \times f_nt^n$$

Differentiating with respect to t,

$$knp(q+pt)^{n-1} + kn(n-1)p^2t(q+pt)^{n-2}$$
$$= 0 + 1^2f_1 + 2^2f_2t + \ldots n^2f_nt^{n-1}$$

Putting $t = 1$

$$knp + kn(n-1)p^2 = \sum_0^n r^2 f_r$$

= sum of the square of the deviations from the origin.

The square of the root-mean-square deviation of the number of successes about the origin

$$= \frac{\Sigma r^2 f_r}{k}$$
$$= np + n(n-1)p^2$$
$$= s^2 \text{ in the notation of Chapter 15 of } Elementary \ Statistics$$

Using the same notation

$$b^2 = (np)^2$$

Therefore,   the variance $= \sigma^2$

$$= s^2 - b^2$$
$$= np + n(n-1)p^2 - (np)^2$$
$$= np - np^2$$
$$= np(1-p)$$
$$= npq$$

Therefore $\sigma = \sqrt{(npq)}$

Alternative Method of Calculating the Arithmetic Mean and Standard Deviation. Since $_nC_r q^{n-r} p^r$ is the probability of obtaining r successes, then in k trials we should expect the frequency of r successes to be $k_n C_r q^{n-r} p^r$.

Arranging the work in the usual way, we have

Number of times success occurs	Frequency	Deviation from Working Mean, 0		
	f	d	fd	fd^2
0	$k_nC_0q^n$	0	0	0
1	$k_nC_1q^{n-1} \times p$	1	$k1_nC_1q^{n-1}p$	$k1^2{}_nC_1q^{n-1}p$
2	$k_nC_2q^{n-2}p^2$	2	$k2_nC_2q^{n-2}p^2$	$k2^2{}_nC_2q^{n-2}p^2$
.	.	.	.	.
.	.	.	.	.
.	.	.	.	.
r	$k_nC_rq^{n-r}p^r$	r	$kr_nC_rq^{n-r}p^r$	$kr^2{}_nC_rq^{n-r}p^r$
.	.	.	.	.
.	.	.	.	.
.	.	.	.	.
n	$k_nC_nq^{n-n}p^n$	n	$kn_nC_np^n$	$kn^2{}_nC_np^n$
Total	k		knp	$knp[1+(n-1)p]$

METHOD

$$\Sigma f = k[{}_nC_0q^n + {}_nC_1q^{n-1}p + \ldots {}_nC_rq^{n-r}p^r + \ldots {}_nC_np^n]$$
$$= k(q+p)^n = k$$
$$\Sigma fd = k[{}_nC_1q^{n-1}p + 2{}_nC_2q^{n-2}p^2 + \ldots r{}_nC_rq^{n-r}p^r + \ldots n{}_nC_np^n]$$

$$= k\left[nq^{n-1}p + \frac{2n(n-1)}{1\times2}q^{n-2}p^2 + \ldots np^n\right]$$

Take a common factor np outside the bracket and reduce each coefficient.

$$\Sigma fd = knp[q^{n-1} + (n-1)q^{n-2}p + \ldots p^{n-1}]$$

The expression inside the bracket is the binomial exansion of $(q+p)^{n-1}$

Therefore

$$\Sigma fd = knp(q+p)^{n-1}$$
$$= knp \qquad [(q+p)^{n-1} = 1]$$
$$\Sigma fd^2 = k{}_nC_1q^{n-1}p + k2^2{}_nC_2q^{n-2}p^2 + \ldots kn^2p^n \qquad ({}_nC_n = 1)$$
$$= kp[nq^{n-1} + 2n(n-1)q^{n-2}p^2 + \ldots n^2p^{n-1}]$$
$$= knp[q^{n-1} + 2(n-1)q^{n-2}p + \ldots np^{n-1}]$$

The right hand side of the equation can be written as the sum of two separate expressions,

$$\Sigma fd^2 = knp[q^{n-1} + (n-1)q^{n-2}p + \ldots p^{n-1}]$$

$$+ knp\left[(n-1)q^{n-2}\times p + \frac{2(n-1)(n-2)}{1\times2}q^{n-3}p^2 + \ldots (n-1)p^{n-1}\right]$$

The first bracketed expression is the expansion of $(q+p)^{n-1}$ and is equal to 1. Take the factors $(n-1)p$ outside the second bracketed expression and cancel the coefficients. The remaining expression in the bracket is the expansion of $(q+p)^{n-2}$ and is equal to unity.

Therefore
$$\Sigma fd^2 = knp + k\times n\times(n-1)\times p^2$$
$$= knp[1+(n-1)p]$$

The arithmetic mean = working mean + average of the deviations from the working mean

$$= 0 + \frac{\Sigma fd}{k}$$

$$= 0 + \frac{knp}{k} = np$$

The square of the root-mean-square deviation from the working mean

$$= s^2$$
$$= knp[1+(n-1)p]/k$$
$$= np[1+(n-1)p]$$

The difference between the working mean and the arithmetic mean = $b = np$.

Therefore
$$\sigma^2 = s^2 - b^2$$
$$= np[1+(n-1)p] - n^2 p^2$$
$$= np(1-p)$$
$$= npq$$

Therefore
$$\sigma = \sqrt{(npq)}$$

CHAPTER 11

The Poisson Distribution

Introduction

A series frequently used in algebra is

$$1+\frac{1}{1!}+\frac{1}{2!}+\frac{1}{3!}+ \ldots \frac{1}{r!}$$

The series can be proved convergent and the limit of its sum as $r \to \infty$ (r tends towards infinity) is denoted by the letter e. That is

$$e = 1+\sum_{1}^{\infty}\frac{1}{r!}$$

The value of e can be found to any degree of accuracy by taking a sufficient number of terms. Thus

$$
\begin{aligned}
1 &= 1 \\
1/1! &= 1 \\
1/2! &= 0\cdot5 & \text{(Dividing 1 by 2)} \\
1/3! &= 0\cdot166667 & \text{(Dividing } 0\cdot5 \text{ by 3)} \\
1/4! &= 0\cdot041667 & \text{(Dividing } 0\cdot166667 \text{ by 4)} \\
1/5! &= 0\cdot008333 \\
1/6! &= 0\cdot001400 \\
1/7! &= 0\cdot000200 \\
1/8! &= 0\cdot000025 \\
1/9! &= 0\cdot000003 \\
1/10! &= 0\cdot000000 \\
\text{Total} &= \overline{2\cdot7183}
\end{aligned}
$$

Therefore the value of e to four decimal places is $2\cdot7183$. The series is known as the exponential series, and the exponential theorem states:

$$e^x = 1+\frac{x}{1!}+\frac{x^2}{2!}+ \ldots +\frac{x^r}{r!}+ \ldots$$

for all values of x. The proof of this theorem can be found in any element-ary algebra book dealing with exponentials. The value of e^x can be calcu-lated from the series or by raising e to the power x by logarithms. e^x occurs so often that tables have been drawn up giving the value of e^x for different values of x. These can be found in practically any set of mathe-matical tables. When using this series in statistics we are required to calculate the value of individual terms.

EXAMPLE

Calculate the value of the fifth term of the series e^x if $x = 1 \cdot 572$.

SOLUTION

$$\text{The fifth term} = x^4/4! = x^4/24$$

Therefore,

$$\log(\text{fifth term}) = 4 \log x - \log 24$$
$$= 4 \times 0 \cdot 1965 - 1 \cdot 3802$$
$$= \bar{1} \cdot 4058$$

Therefore,

$$\text{fifth term} = 0 \cdot 255$$

If r is large, finding $r!$ can be tedious using ordinary mathematical tables. To avoid this, use tables of the logarithms of $r!$ such as those in the *Lanchester Short Statistical Tables*.

EXAMPLE

Calculate the value of e^x when $x = 1 \cdot 48$.

SOLUTION

The simplest way is to use tables of e^x where we find $e^{1 \cdot 48} = 4 \cdot 393$. If tables of e^x are not available, we can use logarithms.

$$\log e^{1 \cdot 48} = 1 \cdot 48 \times \log e$$
$$= 1 \cdot 48 \times 0 \cdot 4343$$
$$= 0 \cdot 6428$$

Therefore,

$$e^{1 \cdot 48} = 4 \cdot 39$$

EXAMPLE

Calculate the value of e^x where $x = -1 \cdot 48$.

SOLUTION

Again the simplest way is to use tables of e^x where we find $e^{-1 \cdot 48} = 0 \cdot 228$. If logarithms are used it is better to write

$$e^{-1 \cdot 48} = 1/e^{1 \cdot 48}$$

Therefore
$$\log e^{-1 \cdot 48} = \log 1 - 1 \cdot 48 \times \log e$$
$$= 0 - 0 \cdot 6428$$
$$= \bar{1} \cdot 3572$$

and
$$e^{-1 \cdot 48} = 0 \cdot 228$$

Alternatively the value of $e^{1 \cdot 48}$ can be found and reciprocal tables used to find $e^{-1 \cdot 48}$

The Poisson Distribution

If in the binomial distribution n is large and p is small so that np remains constant and finite, it can be proved that the binomial distribution reduces to

$$e^{-a}\left[1 + \frac{a}{1!} + \frac{a^2}{2!} + \ldots + \frac{a^r}{r!} + \ldots\right] \qquad \text{(See Mathematical Note 11.1.)}$$

where a equals np, equals the arithmetic mean of the distribution. In this series the probability of exactly r successes is

$$P(r) = e^{-a} \times a^r/r!$$

and the probability of certainty is

$$P(0) + P(1) + \ldots + P(r) + \ldots$$
$$= (e^{-a}) + (e^{-a})\frac{a}{1!} + (e^{-a})\frac{a^2}{2!} + \ldots + (e^{-a})\frac{a^r}{r!} + \ldots$$
$$= e^{-a}\left[1 + \frac{a}{1!} + \frac{a^2}{2!} + \ldots + \frac{a^r}{r!} + \ldots\right]$$
$$= e^{-a} \times e^a$$
$$= e^0$$
$$= 1$$

which we expect.

Relation between the Terms of the Poisson Distribution

Now $$P(r+1) = e^{-a} \times \frac{a^{r+1}}{(r+1)!} \quad \text{and} \quad P(r) = e^{-a} \times \frac{a^r}{r!}$$

Therefore, $$\frac{P(r+1)}{P(r)} = \frac{e^{-a} \times \dfrac{a^{r+1}}{(r+1)!}}{e^{-a} \times \dfrac{a^r}{r!}}$$

$$= \frac{a}{r+1}$$

Therefore, $$P(r+1) = \frac{a}{r+1} P(r)$$

From this it is obvious that $P(r+1)>P(r)$ if $a>(r+1)$, $P(r+1)<P(r)$ if $a<(r+1)$, and $P(r+1) = P(r)$ if $a = (r+1)$.

Arithmetic Mean and Variance

The mean and variance of the binomial distribution is np and npq, respectively. Since the Poisson distribution is the binomial distribution when p is small and therefore q nearly equals unity, we have the variance npq is nearly equal to np. That is, in a Poisson distribution the arithmetic mean, a, nearly equals the variance.

Tabulation of a Poisson Distribution

Since the Poisson distribution only involves one constant, a, it is easier to tabulate than the binomial distribution. Thus if a equals 0.5, the probabilities can be calculated as follows:

$$P(r+1) = \frac{a}{r+1} P(r)$$

Therefore, $$\log P(r+1) = \log a - \log(r+1) + \log P(r)$$
$$\log P(0) = \log e^{-a} = \log e^{-0.5} = 1.7828$$
$$P(0) = 0.6064$$

As in the case of the binomial distribution it is better to arrange the work in the form of a table.

Probability	r	log a	log(r+1)	log P(r+1)	Probability
P(0)				$\bar{1}$·7828	0·6064
P(1)	0	$\bar{1}$·6990	0·0000	$\bar{1}$·4819	0·3037
P(2)	1	$\bar{1}$·6990	0·3010	$\bar{2}$·8799	0·0759
P(3)	2	$\bar{1}$·6990	0·4771	$\bar{2}$·1018	0·0126
P(4)	3	$\bar{1}$·6990	0·6021	$\bar{3}$·1987	0·0016
P(5)	4	$\bar{1}$·6990	0·6990	$\bar{4}$·1987	0·0002
				Total	1·0004

Thus to obtain $\log P(1)$, add $\bar{1}$·6990, subtract 0·0000, and add $\bar{1}$·7828, and this gives $\bar{1}$·4819. To obtain, say, $\log P(3)$, add $\bar{1}$·6990, subtract 0·4771, and add $\bar{2}$·8799, and this gives $\bar{2}$·1018. Theoretically the number of probabilities should be infinite, but the chances of obtaining more than five 'successes' become so small that it is impossible to calculate their probabilities with four-figure logarithms.

If the event was repeated 1000 times the theoretical frequencies of 'successful' events would be:

Successes	0	1	2	3	4	5	
Frequency	606	304	76	13	2	0	Total 1001

[The *Biometrika Tables for Statisticians, Vol 1*, Table 39 gives the probabilities for various values of *a* correct to six decimal places, and the use of these tables saves a large amount of numerical work.]

Comparison with the Binomial Distribution

If in the binomial distribution we put $p = 0\cdot1$ and $n = 5$, then the mean $np = 0\cdot5$. In Table 10.2 the frequencies for the binomial distribution $1000(0\cdot9 + 0\cdot1)^5$ are given, and it is interesting to compare them with the Poisson frequencies above. In Table 11.1 the binomial frequencies for a constant value of $p = 0\cdot1$ and different values of *n* are compared with the Poisson frequencies of distributions with the same means. It is obvious that as *n* increases and *p* decreases the agreement between the frequencies improves.

Table 11.2 compares the frequencies in 1000 trials of the binomial distributions of constant mean 0·5 for different values of *p* and *n* with the Poisson distribution with the same mean.

It is seen that the Poisson distribution is only a good approximation for the binomial distribution if *n* is large and *p* is small, and the greater the value of *n* and the smaller the value of *p* the better the agreement.

Table 11.1 *Comparison of the Frequencies obtained by the Poisson and Binomial Distributions in 1000 Trials*

Successes	Poisson	Binomial	Poisson	Binomial	Poisson	Binomial	Poisson	Binomial	Poisson	Binomial
$p =$		0·1		0·1		0·1		0·1		0·1
$n =$		5		20		50		100		200
$a = $	0·5		2		5		10		20	
0	606	590	135	122	7	5				
1	304	328	271	270	34	29				
2	76	73	271	285	84	78	2	2		
3	13	8	180	190	141	138	8	6		
4	2		90	90	176	181	19	16		
5			36	32	176	185	38	34		
6			12	9	146	154	63	60		
7			3	2	105	108	90	89	1	
8			1		65	66	113	115	1	1
9					36	34	125	130	3	2
10					18	15	125	132	6	5
11					8	6	114	120	11	9
12					3		95	99	18	15
13					1		73	74	27	25
14							52	51	39	36
15							35	33	52	50
16							22	19	64	64
17							13	11	76	77
18							7	5	84	88
19							4	3	89	93
20							2	1	89	94
21							1		84	89
22									77	81
23									67	69
24									56	57
25									45	44
26									34	33
27									25	24
28									18	16
29									12	11
30									8	7
31									5	4
32									3	2
33									2	1
34									1	1
35									1	
Total	1001	999	999	1000	1000	999	1001	1000	998	998

Table 11.2   *Comparison of the Frequencies obtained by the Poisson and Binomial Distributions of Mean 0·5 in 1000 Trials*

Successes	Poisson	Binomial						
	$a = 0.5$	$p = 0.1$ $n = 5$	0·05 10	0·025 20	0·01667 30	0·0125 40	0·01 50	0·001 500
0	606	590	599	603	604	604	606	606
1	304	328	315	309	308	306	305	304
2	76	73	75	75	76	76	76	76
3	13	8	10	12	12	12	12	13
4	2		1	1	1	1	1	2
Total	1001	999	1000	1000	1001	999	1000	1001

Merit of the Poisson Distribution

The merit of the Poisson distribution is its simplicity and its use in isolated or rare events. It plays a large part in industrial sampling and in demographic problems, and has several important applications in biology, traffic control and telephone control, and can be modified for application to particular problems. It justifies its application by the results it produces.

EXAMPLE

A manufacturer produces articles and in spite of all precautions there is a probability that one article in every hundred is defective. The articles are packed in boxes of 500 and the manufacturer wishes to estimate the chance of getting six or more faulty articles in a package.

SOLUTION

Assuming that every package of 500 is a random sample, this becomes a case where the Poisson distribution applies. $p = 0.01$; $n = 500$. Therefore $a = np = 5 =$ variance. It is necessary to calculate the terms of the expansion

$$e^{-5}\left(1 + \frac{5}{1!} + \frac{5^2}{2!} + \cdots\right)$$

and doing this we get,

Number of faulty articles in a package	0	1	2	3	4	5
Probability	0·0067	0·0337	0·0842	0·1404	0·1755	0·1755

Therefore the probability of obtaining 0 to 5 faulty articles is

$$0\cdot0067+0\cdot0337+0\cdot0842+0\cdot1404+0\cdot1755+0\cdot1755 = 0\cdot6160$$

and therefore the probability of obtaining six or more faulty articles is $1-0\cdot6160$ or $0\cdot384$. The notation for the probability that a variable x will be greater than or equal to a given value c, is $P(x\geqslant c)$. In the example $c = 6$, therefore $P(x\geqslant6) = 0\cdot384$.

Poisson Distribution Chart

The problem of finding $P(x\geqslant c)$ is so common that a chart has been drawn giving the probabilities for different values of a and c. The chart is given in Diagram 11.1.

To use the chart proceed as follows:

(i) Having calculated the value of a, find its position on the horizontal scale marked a.

(ii) Follow the vertical at this point until it cuts the curve numbered c.

(iii) Move horizontally to the vertical scale marked P and the reading is the probability for $x\geqslant c$.

In the last example $a = 5$, and this cuts the curve $c = 6$ where the probability reading is $0\cdot38$.

Other uses of the Chart

EXAMPLE

Suppose in the last example the manufacturer decided that he would be satisfied if not more than 5% of the packages contained 20 or more faulty articles. At what level should the average rate of faulty articles be kept during the manufacturing process?

SOLUTION

The problem states that in a random sample of 500 articles the probability of finding 20 or more faulty articles must not be greater than 5%, or 0·05. On the vertical P scale of the chart find $P = 0\cdot05$, and draw a horizontal line to cut the curve numbered 20. At this point draw a vertical line to cut the horizontal scale a and read the value of a at this point. It is $13\cdot25$. Now $a = np$, where $n = 500$.

Therefore, $p = 13\cdot25/500 = 0\cdot0265$.

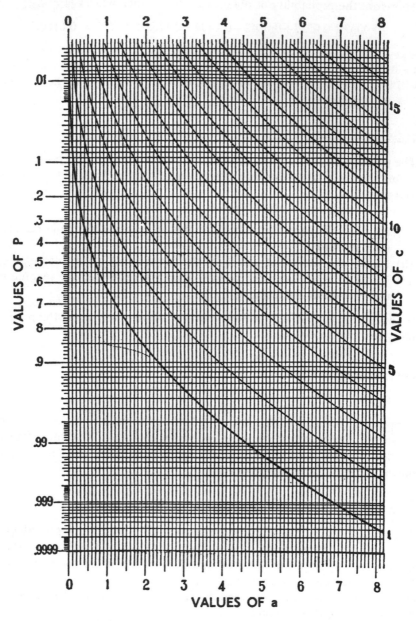

DIAGRAM 11.1 Poisson Distribution Chart

That is, the average rate of faulty articles in manufacture should be 265 in every 10 000, or between two and three in every 100.

Further uses of the Poisson Chart

Suppose we are dealing with a Poisson type distribution for which we know the value of the expectation, a. From the chart we can read off the probability that the event occurs at least 1, 2, 3, and so on, times. For example, suppose $a = 5$; then, running our eye up the vertical line at $a = 5$, we find, by reading the value of the probability opposite the various points where the vertical line cuts the various c curves, that

the event occurs at least once with a probability of 0·993
the event occurs at least twice with a probability of 0·959
the event occurs at least thrice with a probability of 0·875
the event occurs at least four times with a probability of 0·735
the event occurs at least fiive times with a probability of 0·560
the event occurs at least six times with a probability of 0·380
and so on.

Knowing these values, we can obtain the probabilities that the event will occur exactly 0, 1, 2, 3 etc. times. Thus the event occurs at least once with a probability of 0·993. Therefore it will occur 0 times with a probability of $1 - 0·993 = 0·007$. Again, the probability that the event occurs at least twice is 0·959; therefore it must occur less than twice with a probability of $(1 - 0·959) = 0·04$, and we have just found it occurs 0 times with a probability 0·007. Therefore it must occur exactly once with a probability $0·04 - 0·007 = 0·033$. Now,

$$0·04 - 0·007 = (1 - 0·959) - (1 - 0·993) = 0·993 - 0·959 = 0·034$$

In a like manner we can calculate the probabilities of the event occurring other exact numbers of times. Arranging the work in the form of a table, we have:

the event occurs 0 times with a probability of $1·000 - 0·993 = 0·007$
the event occurs 1 time with a probability of $0·993 - 0·959 = 0·034$
the event occurs 2 times with a probability of $0·959 - 0·875 = 0·084$
the event occurs 3 times with a probability of $0·875 - 0·735 = 0·140$
the event occurs 4 times with a probability of $0·735 - 0·560 = 0·175$
the event occurs 5 times with a probability of $0·560 - 0·380 = 0·180$
and so on.

A further use of the chart is to test whether any observed distribution could be described by the Poisson distribution.

EXAMPLE

Consider the following classical data:

The number of deaths caused by the kick of a horse in ten army corps per army corps per annum over twenty years.

Deaths	0	1	2	3	4	
Frequency	109	65	22	3	1	Total = 200

(Data collected by Quetelet and von Bortkiewicz from Prussian army corps, 1875–1894)

The total number of deaths is 122, and the arithmetic mean number of deaths per army corps per annum is 0·61. The probability of each number of deaths is found by dividing the frequencies for each group by the total number of readings, that is 200. Thus we get:

Deaths	0	1	2	3	4
Probability	0·545	0·325	0·11	0·015	0·005

From this table we have,

the probability of less than 1 death occurring   = 0·545
the probability of less than 2 deaths occurring $0·545 + 0·325$ = 0·870
the probability of less than 3 deaths occurring $0·870 + 0·11$ = 0·980
the probability of less than 4 deaths occurring $0·980 + 0·015$ = 0·995

from which we deduce that:

at least 1 death occurs with a probability $1 - 0·545$ = 0·455
at least 2 deaths occur with a probability $1 - 0·870$ = 0·130
at least 3 deaths occur with a probability $1 - 0·980$ = 0·020
at least 4 deaths occur with a probability $1 - 0·995$ = 0·005

If these probabilities are plotted on the Poisson chart, placing each point on its proper curve, $c = 1, 2, 3$, or 4, it will be found they lie on the vertical straight line at the value $a = 0·6$, proving that the distribution follows the Poisson distribution with mean a of 0·6.

EXAMPLE

It is found that in books from a certain printer one in every thousand words has a misprint. Estimate the probability that one hundred words chosen at random from one of his books will have one or more misprints.

SOLUTION

The chance of getting a misprint is $1/1000$. Therefore the average number expected in 100 words is 1/10. The probability of there being no misprints is $e^{-(1/10)}$, that is 0·9048. Therefore the probability of there being one or more misprints is $1 - 0·9048 = 0·0952$.

EXAMPLE

The number of miners killed in Great Britain and the total number employed for the years 1956 to 1961 are given in the table.

Year	1956	1957	1958	1959	1960	1961
Number killed	328	396	327	348	317	235
Number employed (thousands)	781	790	778	756	691	661

(A)

Estimate the probability that a mine employing 1400 miners will be free of fatal accidents in a year. [Assume deaths are randomly distributed.]

SOLUTION

Total killed equals 1951 and the total employed equals 4 457 000. Therefore the probability of being killed = 1951/4 457 000 = 0·000 438. Number employed in the mine = 1400. Therefore the average number expected to receive a fatal accident in a mine employing 1400 men is $1400 \times 0·000\ 438 = 0·6128$. From the chart the probability of one or more deaths when $a = 0·6128$ is 0·45. Therefore the probability of no deaths is 0·55.

Exercise 11.1

1 In a town one handicapped child is born in every 94 births. In a certain month there are 724 births. Find (i) by calculation, (ii) by means of the chart, the chances that in this month ten or more handicapped children are born.

2 The number of vehicles passing a given point in a road was observed for 200 minutes and the frequencies were as follows:

Number of vehicles per min.	0	1	2	3	4	5	6	7	8 or more
Frequency	26	5	52	38	16	8	2	1	1

Use the Poisson chart to test whether these frequencies may be described by a Poisson distribution, and if so, calculate the theoretical frequencies.

3 Of certain manufactured goods 0·2 per cent of the total is defective. If the articles are packed in cartons containing 200 articles, what proportion of cartons is free of defective goods, and what proportion contains two or more defective goods?

4   Calculate the mean and variance of the following distribution. Does it approximate to a Poisson distribution? If you think it does, calculate the theoretical frequencies.

Number of defectives	0	1	2	3	4	5	
Frequency	16	27	23	20	10	4	Total 100

5   The arithmetic mean number of calls received in ten minutes at a firm's switchboard is 5. Assuming Poisson's distribution applies, calculate the probability that in ten minutes (i) 10, (ii) more than 6, (iii) less than 4 calls are received.

6   In a certain building, on the average four electric bulbs have to be replaced every month, and the stock of bulbs is replenished every month. What number of bulbs should be kept in stock to reduce the chance of running short of bulbs to less than 0·01?
    Verify your answer by means of the chart.

7   In two hours, 3600 particles are radiated by radio-active material. Find for the two hours the expected number of ten-second intervals in which 0, 1, 2, etc. particles are radiated. The radiations are random and the radiating material is not appreciably altered.

8   On the average during a rush hour a telephone switchboard handles 282 calls. The maximum number of connections that can be made per minute on the board is 10. Use the Poisson chart to estimate the probability that the board will be overtaxed during any given minute.

9   A business firm finds that on the average two faults occur in their telephone system a year. Use the Poisson chart to find the chance of (i) four, (ii) five or more faults occurring in a year.
    Verify your answers by calculation.

10   A machine manufacturer gives a guarantee of one year and finds that during this period 1 in 550 machines requires adjustment. In a given year he sells 1525 machines. Calculate the probability of there being exactly 0, 1, 2, 3, 4, 5 machines requiring adjustment during the guarantee period.

11   On the average a spare part is used in a factory five times a week. If stocks are replenished once a week, how many spare parts should be kept in stock at the beginning of each week so that the chances are that the stock will be insufficient not more than once every two years?

12 The following erythrocyte count was obtained by means of a haemocytometer. Fit a Poisson distribution to the data and calculate the theoretical frequencies.

Number of Erythrocytes per Square from a Single Blood Specimen in each of 400 Squares in a Graduated Counting Chamber

Number of erythrocytes	0	1	2	3	4	5	6	7	8	9	10	11	12 or more
Number of squares	2	10	21	52	63	77	62	46	41	11	8	5	2 Total 400

13 The number of fatal accidents in U.K. airways for the years 1957 to 1961 were:

Year	1957	1958	1959	1960	1961
Fatal accidents	3	1	0	0	2

(A)

Calculate the probability of there being 0, 1, 2, 3, 4 or more accidents in one of the years.

14 The total number of deaths from road accidents for the years 1956 to 1961 in Great Britain and the total population for these years are given in the table.

Year	1956	1957	1958	1959	1960	1961
Total deaths	5367	5550	5970	6520	6970	6908
Population (thousands)	49811	50058	50278	50577	50963	51350

(A)

Assuming the probability of any individual getting killed in a road accident is the same all over Great Britain, estimate the probability that in any one year a village of 1400 inhabitants will be free of road deaths.

15 Assuming the number of railway servants killed in train accidents in a year follows the Poisson distribution, use the figures given in the table to calculate the probability of there being 1 or more, 2 or more, etc. deaths in a year.

Year	1952	1953	1954	1955	1956	1957	1958	1959	1960	1961
Number of deaths	9	7	1	8	3	4	5	8	6	10

16 Calculate the mean and variance of the following distribution and, if you are satisfied the distribution approximately follows the Poisson Law, use the figures given to calculate the probability of there being 1 or more, 2 or more, etc. deaths in a year.

Year	1951	1952	1953	1954	1955	1956
Deaths in metalliferous mines	7	9	9	6	7	6
	1957	1958	1959	1960	1961	
	9	11	4	2	4	

17   Telephone calls coming in to a switchboard follow a Poisson distribution with a mean 3 per minute. Find the probability that in a given minute there will be five or more calls.

If the duration of every call is three minutes and at most twelve calls can be connected simultaneously, find an approximation to the probability that at a given instant the switchboard is fully loaded.   (AEB)

18   Show that for the Poisson distribution in which the probabilities of 0, 1, 2, . . . r, . . . successes are respectively

$$e^{-m}, \qquad me^{-m}, \qquad \frac{m^2}{2!} e^{-m} . . ., \qquad \frac{m^r}{r!} e^{-m}, . . .$$

the mean and variance of the number of successes are each equal to m.

In a factory it was observed that on 24 days there was no breakdown, on 34 days there was one breakdown, on 25 days there were two breakdowns, on 11 days there were four breakdowns. Find the average number of breakdowns per day.

Compare the observed frequencies with the frequencies of a Poisson distribution with the same total frequency and mean.

Comment on the comparison.   (AEB)

19   On the average a school wins two University Scholarships per year. By means of a Poisson probability chart, or otherwise, find the probability of the school winning (i) 4 or more scholarships in one year, (ii) exactly 4 scholarships in one year.   (AEB)

20   It is found by experiment that the distribution of values of a certain variate is given by Poisson's formula. Thus, if the mean value of the variate is μ, the proportion of occasions on which the value r is recorded is equal to the coefficient of t^r in the expansion of $e^{\mu(t-1)}$ in ascending powers of t. Discuss carefully the circumstances in which Poisson's distribution is a close approximation to a binomial distribution.

Certain mass-produced articles, of which 1 in 200 is defective are packed in cartons each containing 100 articles. Assuming that the number of defective articles per carton follows Poisson's distribution, find the probability that a carton selected at random will contain (a) no defective article, (b) two or more defective articles.   (C)

21   The following table shows the results of recording the telephone calls

handled at a village telephone exchange between 1 p.m. and 2 p.m. on each of 100 weekdays (e.g. on 36 days no such calls were made):

Calls	0	1	2	3	4 or more
Days	36	35	22	7	0

Assuming that calls arrive independently and at random, estimate

(i) the mean m of the corresponding Poisson probability distribution;
(ii) the probability that if the operator is absent for 10 minutes, no call will be missed;
(iii) the probability that if the operator is absent for 10 minutes, two or more calls will be missed.

(Answers need only be given to an accuracy of three significant figures.)

(C)

22   The number of accidents per shift in a factory in 200 shifts is shown in the following frequency table:

Accidents per shift	0	1	2	3	4	5	6	Total
Frequency	130	51	12	4	2	1	0	200

Calculate the mean number of accidents per shift.

Use the Poisson distribution with this mean to calculate the chance of three or more accidents occurring in a single shift. ($e = 2 \cdot 718$.) (O and C)

23   Explain briefly what is meant by the Poisson distribution of rare events and its relation to the binomial distribution. Prove that the mean of the distribution is equal to its variance.

A shopkeeper's sales of washing machines are four per month on the average. Assuming that the monthly sales fit a Poisson distribution, find to what number he should make up his stock at the beginning of each month so that his chance of running out of machines during the month will be less than 4 per cent. (O and C)

24   Explain what is meant by the Poisson distribution for events of rare frequency and show that the mean of the distribution is equal to its variance.

Samples of 40 articles at a time are taken periodically from the continuous production of a machine and the number of samples containing 0, 1, 2, . . . defective articles are recorded in the following table:

No. of defectives per sample	0	1	2	3	4	5	6	Total
No. of samples	30	23	27	14	4	2	0	100

Find the mean number of defectives per sample.

Assuming that this is the mean of the population and that the Poisson distribution applies, find the chance of: (a) a sample containing 4 or more defectives, (b) two successive samples containing between them 4 or more defectives.   (O and C)

25   State what is meant by (i) the binomial distribution, (ii) the Poisson distribution, explaining clearly the symbols used. Give the mean and variance of each distribution.

In an examination 60 per cent of the candidates pass but only 4 per cent obtain distinction. Use the binomial distribution to calculate the chance that a random group of 10 candidates should contain at most two failures.

Use the Poisson distribution to calculate the chance that a random group of 50 candidates should contain more than one distinction.   (O and C)

26   A process for making electric light bulbs produces, on the average, one bulb defective in 100, when under control. Use the Poisson distribution to calculate the probability that a sample of 50 will contain (i) 0, (ii) 1, or (iii) 2 defective bulbs.

The bulbs are delivered in batches of 1000, from which 50 are taken at random. If at most one of the 50 is found to be defective, the whole batch is accepted. Calculate the probability of accepting a batch containing 50 defective bulbs.   (AEB)

Mathematical Note 11.1.

The arithmetic mean of a binomial distribution was shown to be np and the variance npq. It follows that if np is held constant, equal to a, then as $p \to 0$, $n \to \infty$, and $q \to 1$, the mean becomes a and the variance also $\to a$.

It is an important property of the Poisson distribution that

the variance (σ^2) = the mean (a)

This is used as a first test that observed frequencies obey the Poisson distribution.

To show the Poisson distribution is the limiting form of the binomial distribution when $p \to 0$ we proceed as follows.

The probability generating function of the binomial distribution

$$
\begin{aligned}
&= (q + pt)^n \\
&= [1 + p(t - 1)]^n \qquad \text{(since } q = 1 - p) \\
&= \left[1 + \frac{a(t - 1)}{n}\right]^n \qquad \text{(since } np = a)
\end{aligned}
$$

A well known limit in algebra is

$$\left(1+\frac{x}{n}\right)^n \to e^x \qquad \text{as } n\to\infty$$

Hence if a is a constant and $n\to\infty$ the probability generating function of the binomial distribution tends to the form

$$e^{a(t-1)}$$
$$= e^{-a}e^{at}$$
$$= e^{-a}\left[1+\frac{at}{1!}+\frac{a^2t^2}{2!}+ \ldots +\frac{a^rt^r}{r!}+ \ldots \right]$$

The probability of r 'successes' is the coefficient of t^r, that is

$$e^{-a}a^r/r!$$

Therefore in k trials the theoretical frequency of r 'successes' is

$$ke^{-a}a^r/r!$$

If f_r is the frequency of r successes, the generating function can be written as

$$f_0+f_1t+f_2t^2+ \ldots = ke^{-a}e^{at} \qquad (11.1)$$

Differentiating with respect to t, we have

$$f_1+2f_2t+3f_3t^2+ \ldots = kae^{-a}e^{at} \qquad (11.2)$$

Multiplying both sides by t and differentiating again

$$f_1+2^2f_2t+3^2f_3t+ \ldots = kae^{-a}e^{at}+ka^2e^{-a}te^{at} \qquad (11.3)$$

Putting $t = 1$ in Equations (11.1) and (11.2), we have

$$\Sigma f = k \quad \text{and} \quad \Sigma rf = ka$$

Therefore,   the mean $= \dfrac{\Sigma rf}{\Sigma f} = a$

Putting $t = 1$ in Equation (11.3), we have

$$\Sigma r^2 f = ka+ka^2$$
$$= \text{sum of the squares of the deviations from the origin}$$

The square of the root-mean-square deviation of the number of 'successes' about the origin

$$= \frac{\Sigma r^2 f}{\Sigma f}$$
$$= k(a+a^2)/k$$
$$= a+a^2$$
$$= s^2$$

Now
$$b^2 = a^2$$

Therefore, the variance

$$\sigma^2 = s^2 - b^2$$
$$= a+a^2-a^2$$
$$= a$$

CHAPTER 12

The Normal Distribution

Difficulties connected with Continuous Variables

When dealing with frequency distributions with continuous variables, we found it leads to simplification if the mass of figures is divided into classes. To represent this frequency distribution diagrammatically we draw a histogram by drawing lengths proportionate to the class-interval along the horizontal axis and on each length constructing a rectangle whose area is proportionate to the class-frequency for that class-interval. This representation, although helpful up to a point, is not entirely satisfactory. For instance, it suggests that the items are spread evenly over the class-interval, but this is very seldom so. Another method suggested is to join up successive mid-points of the tops of the rectangles and form a frequency polygon. This again is open to the objection that the area above the class-interval only approximates to the class-frequency. Also these diagrams can sometimes give a false impression of frequency and frequency changes in the population as a whole, of which the observations are but a sample. We seldom, if ever, deal with the whole population and it is only within certain broad limits that the observations can give us information which is truly representative of the whole population. Even if we have an adequate sample, and it is unbiased, the numbers obtained in each class are still subject to random sampling errors, and these errors must be calculated before we can lay down the probable limits within which our sample may be regarded as representative of the population as a whole.

Overcoming the Difficulties

These difficulties disappear if we can find a curve to 'fit' the observed statistics. Since the observed statistics are subject to random sampling errors we must not expect the curve to pass through all, or possibly any,

of the observed points. The curve must smooth out the irregularities introduced from ordinary observation.

We could draw a curve by eye, but individuals would disagree on the
curve of best fit, and the method would not be accurate enough. It is
better to find an algebraical formula which gives the relationship between
the variables. It is desirable that the equation should depend on all the
statistics collected, and that the constants involved should be easily
computed from all the given data.

The most common histograms and frequency polygons start with a low
frequency, rise to a maximum, then fall again. The easiest curve to deal
with algebraically is a symmetrical curve. This is not necessarily the
most common shape but fortunately it occurs often enough to justify a
thorough investigation of its properties.

The Normal Curve

The curve
$$y = y_0 \exp(-x^2/2\sigma^2)$$

is such a curve and is known as the normal curve. Its shape is approximated to by the frequency polygon of a binomial distribution when n
becomes large, and the equation can be derived as a limiting form of
the binomial distribution when n tends towards infinity (see Diagram 10.2
and Mathematical Note 12.1). The shape of the curve is illustrated in
Diagram 12.1, and it stretches from $x = -\infty$ to $x = +\infty$. y_0 is the
maximum ordinate and is the value of y when $x = 0$. The standard
deviation of the distribution is σ (see Mathematical Note 12.3).

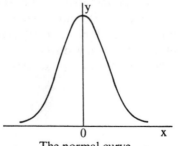

The normal curve

DIAGRAM 12.1

Properties of the Curve

The curve is obviously symmetrical about the y-axis, for plus and minus values of x give the same value of y. The mean, median and mode of the distribution coincide, and are at $x = 0$. When dealing with histograms we stated that the area of the histogram must, on the chosen scale, be equal to the total frequency of the distribution. The same condition applies to the curve, and the area above a given interval of x must equal the total frequency in that interval. The area under the curve can be found by the calculus, and is equal to

$$y_0 \sigma \sqrt{2\pi}$$

If N is the total frequency of the distribution, then

$$N = y_0 \sigma (2\pi)^{\frac{1}{2}}$$

or,

$$y_0 = N/\sigma(2\pi)^{\frac{1}{2}}$$

and the equation of the curve becomes,

$$y = \frac{N}{\sigma(2\pi)^{\frac{1}{2}}} \exp(-x^2/2\sigma^2)$$

The origin is at the mean value of x. If it is necessary to refer the curve to some other point as origin, all we have to do is write $(x-m)$ for x in the above equation, where m is the x-coordinate of the mean; that is, the equation becomes

$$y = \frac{N}{\sigma(2\pi)^{\frac{1}{2}}} \exp[-(x-m)^2/2\sigma^2]$$

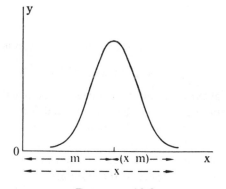

DIAGRAM 12.2

EXAMPLE

Calculate the equation of the normal curve whose mean is 8, standard deviation 3 and total frequency 1000.

SOLUTION

$N = 1000$; $\sigma = 3$; $m = 8$. Therefore, the equation of the normal curve is

$$y = \frac{1000}{3 \times (2\pi)^{\frac{1}{2}}} \exp[-(x-8)^2/18]$$

The value of the ordinate at $x = 11$, is

$$132.9 \exp[-(11-8)^2/18]$$
$$= 132.9 \exp[-(x-8)^2/18]$$
$$= 132.9 \, e^{-\frac{1}{2}}$$
$$= 80.60$$
$$= 81 \qquad \text{to the nearest integer}$$

Exercise 12.1

Calculate the equations of the normal curves which satisfy the following conditions, and the values of the ordinates for the given values of the *x*-variable.

	Mean		Total frequency	Value of *x*-variable		
(i)	12	2	2000	4	10	16
(ii)	3	5	1000	=7	0	13
(iii)	10·3	2·3	1521	6·4	10·5	21·2
(iv)	=4	3	1762	=8	0	4

The Standard Normal Curve

The normal curve is so important that it has been standardised into

$$y = (2\pi)^{-\frac{1}{2}} \exp(-x^2/2)$$

that is, in the general curve N and σ have been made equal to unity and the origin is at the mean. Tables have been computed for this standard curve (Tables 1 and 2) giving the value of y corresponding to a given value of x, and the area enclosed by the curve, the *x*-axis and a given ordinate.

Practice in the use of these tables is essential. The area under the standard curve is unity, and the *x* of the standard curve is $(x-m)/\sigma$ of the general curve. Since the curve is symmetrical about the mean the tables generally give the values of the ordinates and the area under the curve for positive values of x only.

Since there may be some confusion between the x and y of the standard and general curves, X and Y will be used when referring to the standard curve and the lower case x and y when referring to the general curve. Thus the standard curve will be written

$$Y = (2\pi)^{-\frac{1}{2}} \exp(-X^2/2)$$

and the general curve

$$y = (2\pi)^{-\frac{1}{2}} \exp[-(x-m)^2/2\sigma^2]$$

The tables refer to the standard form of the equation and the general equation must be reduced to the standard form before the tables can be used. The relations between X and Y and x and y are

    (i) $X = (x-m)/\sigma$, or $x = m+X\sigma$,
    (ii) $Y = \sigma y/N$, or $y = NY/\sigma$,

and an area under the general curve for a given value of x is N times the corresponding area under the standard curve for the corresponding value of X.

EXAMPLE

If $N = 1000$; $m = 8$; $\sigma = 3$; use tables to find
 (i) the value of the ordinate,
 (ii) the area under the curve to the above ordinate at $x = 11$.

SOLUTION

$$X = (x-m)/\sigma = (11-8)/3 = 1$$

From the table of ordinates (Table 2) we get the value of Y at $X = 1$, which is 0·2420.

Therefore,
$$\begin{aligned} y &= 0·2420N/\sigma \\ &= 0·2420 \times 1000/3 \\ &= 80·67 = 81 \quad \text{to the nearest integer} \end{aligned}$$

From the table of areas (Table 1) we obtain the area of the normal curve up to $X = 1$, which is 0·8413. Therefore, in the general curve the total frequency up to $x = 11$ is

$$1000 \times 0·8413 = 841$$

Further Properties of the Normal Curve

The curve extends to infinity on either side of the origin, and it can be plotted by using the values of the ordinates given in Table 2. Whether the curve looks broad and flat or narrow and high depends on the scale chosen for X and Y, (see Diagram 12.3), but in each case the area above any given interval of X is proportionate to the frequency in that interval, and is equal for all scales.

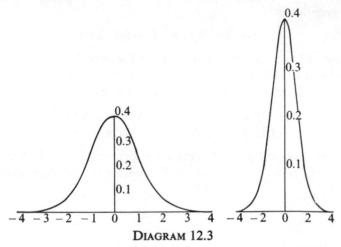

DIAGRAM 12.3

From Table 1 it will be noticed that the area under the curve up to $X = 4$ is 0·99997 units. Therefore, by symmetry, the area up to $X = -4$ is $1-0·99997$ units, and therefore the area between $X = \pm4$ is 0·99997 $-(1-0·99997)$ units, that is, 0·99994 units.

Similarly the area between $X = \pm3$ is $0·99865-(1-0·99865) = 0·99730$, and the area between $X = \pm2$ is $0·9772-(1-0·9772) = 0·9544$, and the area between $X = \pm1$ is $0·8413-(1-0·8413) = 0·6826$.

Therefore, the frequencies in the general curve between the

$$\text{mean}\pm3\sigma = 0·99730N,$$
$$\text{mean}\pm2\sigma = 0·9544N,$$
$$\text{mean}\pm\sigma = 0·6826N,$$

and this is the reason for the statement that in a normal distribution, approximately 0·68 of the items lie in the range (mean$\pm\sigma$), 0·95 of the items in the range (mean$\pm2\sigma$), and nearly all the items in the range (mean$\pm3\sigma$). Also from Table 1 on areas it will be seen that in the standard curve, 0·75 of the area lies to the left of the ordinate at X = 0·675, that is 75% of the frequency will lie to the left of this ordinate which is therefore the upper quartile. Therefore, in the general curve the upper quartile will be at $x = 0·675\sigma$. Similarly the lower quartile will be at $x = -0·675\sigma$. This gives rise to the statement that in a fairly symmetrical one-humped distribution

$$\text{quartile deviation} = (2/3)\text{standard deviation}$$

EXAMPLE

In a normal distribution the mean, standard deviation and total frequency are 11·3, 2·4, and 796, respectively. How many items will

(i) exceed 12·7 in value,
(ii) be less than 9·6 in value,
(iii) lie between 10·5 and 13·2?

SOLUTION

$$N = 796; \quad m = 11·3; \quad \sigma = 2·4$$

(i) $x = 12·7$.   Therefore $X = (x-m)/\sigma = (12·7-11·3)/2·4 = 0·5333$

From Table 1 we find the proportion of the area up to X = 0·5333 is 0·7031. Therefore the proportion of the area exceeding this is 1 − 0·7031 = 0·2969. Therefore the number of items in the distribution exceeding 12·7 in value is 0·2969 × 796 = 236.

(ii) $x = 9·6$.   Therefore, $X = (x-m)/\sigma = (9·6-11·3)/2·4 = -0·7083$

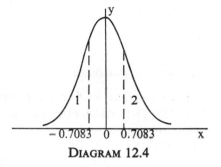

DIAGRAM 12.4

The frequency we are interested in is proportionate to the area marked 1 in Diagram 12.4. We cannot read this area from the tables but, by symmetry, we know it is equal to the area marked 2. From Table 1 the area up to X = 0·7083 is 0·7606. Therefore, area 2 = 1−0·7606 = 0·2394, and area 1 = 0·2394. The number of items represented by this area = 0·2394 × 796 = 191.

(iii) To solve this we must calculate the number of items up to $x = 10·5$ and up to $x = 13·2$ and subtract.

$$\text{If } x = 13·2, \ X = (x-m)/\sigma = (13·2-11·3)/2·4 = 0·7917$$

From Table 1 the number of items up to $x = 13·2$ is proportionate to 0·7869.

$$\text{If } x = 10·5, \ X = (10·5-11·3)/2·4 = -0·3333$$

From Table 1 the proportion of items up to X = 0·3333 is 0·6325. Therefore, the proportion of items beyond X = 0·3333 is 1−0·6325 = 0·3675 and, the proportion of items up to X = −0·3333 must be 0·3675. It follows that the proportion of items between $x = 10·5$ and $x = 13·2$ is

$$0·7869-0·3675 = 0·4194$$

Therefore, the number of items is 0·4194 × 796 = 334.

Exercise 12.2

1   Calculate using Tables 1 and 2 the values of the ordinates and of the frequencies up to the given ordinates in the normal curves of Exercise 12.1.

2   In the normal curve whose frequency is 2000, standard deviation 2, and mean 6, how many items are there whose values are

(i) less than (a) 6·5, (b) 4·2,
(ii) greater than (a) 8·3, (b) 5·1
(iii) between (a) 3·7 and 8·1?

Use of Table 1 for Calculating Probabilities

In the example in the last paragraph we found the number of items that exceeds the value 12·7 is 236. The total number of items is 796. Therefore the probability that the value of an item exceeds 12·7 is 236/796. Now 236 is obtained by multiplying 0·2969 by 796. Therefore the probability that the value of the item exceeds 12·7 is 0·2969, and this is the area under the standard curve proportionate to the number of items. For this reason the standard curve is often called the probability curve.

From the last example it follows immediately that the probability that the value of an item of the distribution chosen at random is less than 9·6 is 0·2394, and the probability that its value lies between 10·5 and 13·2 is 0·4194.

Let us now consider a slightly more difficult example.

EXAMPLE

The following distribution is given in a book by W. A. Shewhart titled *Economic Control of Quality Manufactured Product.*

Distribution of Depth of Sapwood in Telegraph Poles

Depth of Sapwood (cm) Central values	Frequency	Depth of Sapwood (cm) Central values	Frequency
2·50	2	9·25	123
3·25	29	10·00	82
4·00	62	10·75	48
4·75	106	11·50	27
5·50	153	12·25	14
6·25	186	13·00	5
7·00	193	13·75	1
7·75	188		
8·50	151	Total	1370

(i) Calculate the equation of the normal curve with the same mean, standard deviation and total frequency.
(ii) Calculate the theoretical frequencies for the same class-intervals.

SOLUTION

This is a fairly symmetrical one-humped distribution and it is legitimate to use Sheppard's correction. Calculating the mean and corrected standard deviation, we have:

$$\text{mean} = 7·285, \qquad \text{corrected } \sigma = 1·995$$

The normal equation is $y = \dfrac{N}{\sigma(2\pi)^{\frac{1}{2}}} \exp[-(x-m)^2/2\sigma^2]$

$$= \frac{1370}{1·995(2\pi)^{\frac{1}{2}}} \exp \frac{-(x-7·285)^2}{2(1·995)^2}$$

$$= 275·35 \exp \frac{-(x-7·285)^2}{7·96}$$

The ordinate at $x = 8.50$ is calculated by substitution in the equation.

It is
$$y = 275.35 \exp \frac{-(8.50-7.285)^2}{7.96}$$

$$= 275.35 \exp(-0.1854)$$
$$= 275.35 \times 0.8308$$
$$= 228.76$$

Alternatively if we use Table 2 to calculate the ordinate at $x = 8.50$, we proceed as follows:

$$X = (x-m)/\sigma = (8.50-7.285)/1.995 = 0.6090$$

Therefore, $Y = 0.3313$ from Table 2, and

$$y = \frac{N}{\sigma} Y = \frac{1370}{1.995} \times 0.3313 = 227.4.$$

Now 8.50 is the mid-ordinate of the interval 8.125 to 8.875, and the frequency in this class is the area ABCD in Diagram 12.5 where A is at 8.125 and B at 8.875.

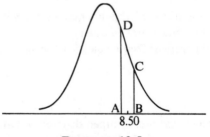

$$8.50$$

DIAGRAM 12.5

If we take ABCD to be a trapezium its area is $228.76 \times 0.75 = 171$ to the nearest unit. Therefore the theoretical frequency in the interval whose central value is 8.5 is 171. A more accurate method is to use Table 1. The area of the class-interval ranges between the ordinates at $x = 8.125$ and $x = 8.875$. If $x = 8.125$; $X = (8.125-7.285)/1.995 = 0.4211$ and the area under the standard curve up to $X = 0.4211$ is, from Table 1, equal to 0.6632. If $x = 8.875$; $X = (8.875-7.285)/1.995 = 0.7970$. The area under the standard curve up to this value of X is 0.7872. Therefore the area proportionate to ABCD is $(0.7872-0.6632) = 0.1240$, and this makes the frequency in the general curve as $0.1240 \times N = 0.1240 \times 1370 = 170$. The frequency for each class-interval can be calculated in the same way. It is best to arrange the work in the form of a table.

Col. 1 Lower interval limit	Col. 2 $X = \dfrac{x-m}{\sigma}$	Col. 3 Area to right of X	Col. 4 Proportion of area in interval	Col. 5 Col. 4 × 1370	Col. 6 Theoretical frequency
$-\infty$	$-\infty$	1·0000	0·01353	18·54	19
2·875	$-2·2105$	0·98647	0·01977	27·08	27
3·625	$-1·8346$	0·9667	0·0389	53·29	53
4·375	$-1·4586$	0·9278	0·0672	92·06	92
5·125	$-1·0827$	0·8606	0·1005	137·69	138
5·875	$-0·7068$	0·7601	0·1305	178·79	179
6·625	$-0·3308$	0·6296	0·1476	202·21	202
7·375	0·0451	0·4820	0·1452	198·92	199
8·125	0·4211	0·3368	0·1240	169·88	170
8·875	0·7970	0·2128	0·0924	126·59	127
9·625	1·1729	0·1204	0·0596	81·65	82
10·375	1·5489	0·0608	0·0337	46·17	46
11·125	1·9248	0·0271	0·0164	22·47	22
11·875	2·3008	0·0107	0·00698	9·56	10
12·625	2·6767	0·00372	0·00257	3·52	4
13·375	3·0526	0·00115	0·00115	1·58	2
∞	∞	0	0	0	0
				Total	1372

METHOD

Col. 1
In the given distribution the first class-interval ranges from 2·125 to 2·875, but in the normal distribution the lower limit is $-\infty$ to 2·875. Therefore, to allow all the normal distribution to be represented, take the first class-interval from $-\infty$ to 2·875.
Col. 2
Calculate X corresponding to the lower limit of each class-interval.
Col. 3
From Table 1 calculate the area to the right of the lower limit of each class-interval.
Col. 4
By subtraction calculate the proportion of the area in each class-interval. Thus, the area of the standardized normal curve to the right of $-\infty$ is 1, and to the right of $X = -2·2105$ is $1 - 0·98647 = 0·01353$. In the class-interval 3·625 to 4·375 the proportion of the area is $0·9667 - 0·9278 = 0·0389$.
Col. 5
Col. 4 is multiplied by the total frequency, 1370, to obtain the area in each class-interval of the given normal curve.

Col. 6

The figures in Col. 5 are written correct to the nearest integer to obtain the frequency in each class-interval. Owing to the rounding off of the numbers the total frequency is 1372.

Exercise 12.3

In the following distributions calculate the

(i) equation of the normal curve with the same mean, standard deviation and total frequency,

(ii) theoretical frequencies for the class-intervals

    (a) by calculating the mid-ordinates of the class-intervals,

    (b) by using Table 1.

1 | x | 10– | 12– | 14– | 16– | 18– | 20– | 22– | 24– | 26– | |
|---|-----|-----|-----|-----|-----|-----|-----|-----|-----|---|
| f | 5 | 31 | 105 | 209 | 271 | 220 | 116 | 38 | 5 | Total = 1000 |

2 | x | 30– | 35– | 40– | 45– | 50– | 55– | 60– | 65– | |
|---|-----|-----|-----|-----|-----|-----|-----|-----|---|
| f | 3 | 31 | 150 | 310 | 325 | 149 | 30 | 2 | Total = 1000 |

3 | x | 190–193 | 194–197 | 198–201 | 202–205 | 206–209 | 210–213 |
|---|---------|---------|---------|---------|---------|---------|
| f | 3 | 5 | 8 | 13 | 20 | 25 |
| | 214–217 | 218–221 | 222–225 | 226–229 | 230–233 | 234–237 |
| | 28 | 37 | 27 | 23 | 19 | 14 |
| | 238–241 | 242–245 | 246–249 | 250–253 | | |
| | 7 | 6 | 3 | 2 | Total = 240 | |

4   Distribution of red blood cell count for 140 men.

Cell count (10^6 per mm^3)	4·6 or less	4·9 or less	5·2 or less
f	5	15	29
	5·5 or less	5·8 or less	6·1 or less
	35	33	15
	6·4 or less	6·7 or less	
	6	2	

The Use of the Normal Curve for a Discrete Variable

When studying the binomial distribution it was shown that if $p = \frac{1}{2}$ or n is large then the frequency polygon of the distribution approximates to a symmetrical distribution similar to the normal curve. This leads us to expect that if a discrete distribution satisfies either, or both, the above conditions, then the properties of the normal curve can be used to solve, with a sufficient degree of accuracy, problems pertaining to the discrete distribution.

EXAMPLE

Two hundred dice are thrown 1000 times. Calculate the number of times exactly forty aces can be expected.

SOLUTION

With the usual notation, $N = 1000$; $p = 1/6$; $q = 5/6$; $n = 200$. $m = np$ $= 200 \times 1/6 = 33.3333$; $\sigma = \sqrt{(npq)} = \sqrt{(200 \times 1/6 \times 5/6)} = 5.2705$. One of the conditions mentioned in the last paragraph is satisfied, that is, n is large.

METHOD 1

By the binomial distribution the number of times exactly 40 aces are to be expected is

$$1000 \,_{200}C_{40}(1/6)^{40}(5/6)^{160} = 32.86 \simeq 33$$

METHOD 2

By calculating the ordinate of the normal curve. If $x = 40$, then

$$X = (40-33.3333)/5.2705 = 1.2649$$

From Table 1 the corresponding value of Y is 0.1794. Therefore,

$$y = Y\frac{N}{\sigma} = 0.1794 \times \frac{1000}{5.2705} = 34.03 = 34 \text{ to the nearest integer.}$$

Since in the binomial distribution 39, 40, 41, etc. are the mid-points of the classes in the histogram or frequency polygon, the interval corresponding to 40 ranges from $x = 39.5$ to 40.5, and so on. *This must always be taken into account when using the continuous normal curve to represent a discrete distribution.* Therefore the area above the interval 39.5 to 40.5 is 34×1. Therefore the frequency of 40 aces is 34.

METHOD 3

Again the interval ranges from 39.5 to 40.5. If $x = 39.5$ then

$$X = (39.5-33.3333)/5.2705 = 1.1700$$

Area up to $X = 1.1700$ is, from Table 1, equal to 0.8790. If $x = 40.5$ then
$$X = (40.5 - 33.3333)/5.2705 = 1.3598$$

Area up to $X = 1.3598$ is 0.9131. Therefore proportionate area in interval is $0.9131 - 0.8790 = 0.0341$. Therefore the frequency of 40 aces is $0.0341 \times 1000 = 34$.

EXAMPLE

Twenty unbiased coins are tossed 1000 times. Calculate the number of times exactly 9 heads are expected.

SOLUTION

$N = 1000$; $n = 20$; $p = \frac{1}{2}$; $q = \frac{1}{2}$; $m = np = 20 \times \frac{1}{2} = 10$; $\sigma = \sqrt{(npq)}$
$= \sqrt{(20 \times \frac{1}{2} \times \frac{1}{2})} = 2.236$.
In this case n is not large, but $p = \frac{1}{2}$.

METHOD 1

By the binomial distribution

$$\text{Frequency} = 1000 \,_{20}C_9 (\tfrac{1}{2})^9 (\tfrac{1}{2})^{11}$$

$$= \frac{1000 \times 20!}{11! \times 9!} \times (\tfrac{1}{2})^{20}$$

$$= 160.2$$
$$= 160 \text{ to the nearest integer}$$

METHOD 2

By calculating the ordinate of the normal curve. $x = 9$. Therefore,

$$X = (9 - 10)/2.236 = -0.4496$$

From Table 2, $Y = 0.3602$. Therefore,

$$y = 1000/2.236 \times 0.3602 = 161.1$$

The width of the interval is unity, and therefore the area above the interval is 161.1, and the frequency to the nearest integer is 161.

METHOD 3

By areas. If $x = 8.5$, then

$$X = (8.5 - 10)/2.236 = -0.6708$$

Area to the right of $X = -0.6708$ is 0.7489. If $x = 9.5$, then

$$X = (9.5 - 10)/2.236 = -0.2236$$

Area to the right of $X = -0.2236$ is 0.5885. Therefore the area of the interval $= 0.7489 - 0.5885 = 0.1604$. Therefore the frequency of nine heads is $0.1604 \times 1000 = 160$.

Advantage of using the Normal Curve

It might appear from the above examples that if tables of logarithms of factorials are available that there is no advantage in using the theory of the normal curve. If the problem is altered slightly the advantage becomes obvious.

EXAMPLE

Twenty unbiased coins are tossed 1000 times. Calculate the number of times 9 or more heads are to be expected.

SOLUTION

By the theory of the binomial distribution the number of heads expected is

$$1000[_{20}C_9 + _{20}C_{10} + _{20}C_{11} + \ldots + _{20}C_{20}] \times (\tfrac{1}{2})^{20}$$

which is a tedious calculation even if logarithms of factorials are available.

If we use the theory of the normal distribution, the probability of obtaining 9 or more heads is proportionate to the area to the right of the appropriate ordinate in the standardized curve, that is to the right of the ordinate $x = 8.5$ in the general curve. If $x = 8.5$, then

$$X = (8.5 - 10)/2.236 = -0.6708 \qquad [m = np = 20 \times \tfrac{1}{2} = 10]$$

Area to the right of $X = -0.6708$ is, by Table 1, 0.7489. Therefore, the frequency of 9 or more heads is 749.

Exercise 12.4

1   A coin is tossed a number of times. Calculate the chance of obtaining,

(i) 52 or more heads in 100 tosses,
(ii) 520 or more heads in 1000 tosses,
(iii) 5200 or more heads in 10 000 tosses.

2   From experience an employer finds he has to reject 30% of applicants for employment. What is the chance that after interviewing 200 applicants he will find at least 80% of them satisfactory?

3   Fifty-one boys are born for every forty-nine girls. If six babies are born on the same day in a hospital, what is the probability of five of them being boys?

4   One hundred and twenty coins are tossed. What is the probability of obtaining (i) 70 or more heads, (ii) less than 50 heads, (iii) at least 65 heads or tails?

Probability Graph Paper

If the cumulative frequency of a normal distribution is plotted against the value of the variable on natural scale graph paper, we obtain the familiar-shaped curve, called the ogive, as in Diagram 12.6. For convenience the cumulative frequency is given as a percentage of the total frequency. From the table of areas of the normal curve we see that 0·8413 or 84·13% of the area lies to the left of the mean plus one standard deviation. Since the mean and median of a normal distribution coincide, the difference between the values of the variable at 84·13% and 50% should give us a measure of the standard deviation. In Diagram 12.6 the values of the variable at these two points are 5 and 4. Therefore the standard deviation is 1. If the ogive is made into a straight line it could be possible to obtain a more accurate estimate of the mean and standard deviation. Probability graph paper has its cumulative percentage frequency scale chosen so that the ogive for a normal distribution becomes a straight line as in Diagram 12.7.

Pads of the paper can be purchased but the percentage lines can be ruled on plain paper with the aid of the normal tables. The relation between the cumulative percentage frequency scale and the natural scale of the variable is as follows. If p is the cumulative percentage frequency corres-

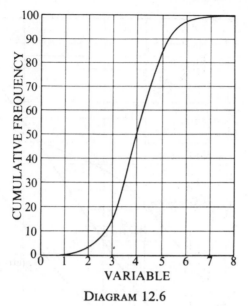

VARIABLE

DIAGRAM 12.6

ponding to a value x of the variable, then the *distance* representing $(p-50)\%$ is proportionate to the distance representing $(x-\bar{x})/\sigma$ on the natural scale.

Skew Distributions

If the cumulative percentage frequencies of moderately skew distributions are plotted on ordinary graph paper, we get a curve very similar to that for a normal distribution, but if they are plotted on probability graph paper we obtain a curve, instead of a straight line, and it is thus easier to see the distribution is skew.

Consider the two following distributions.

Distribution A

Upper class-limit	1	2	3	4	5	6	7	8
Frequency	20	20	30	40	150	350	295	95
Percentage cumulative frequency	2	4	7	11	26	61	90·5	100

Distribution B

Upper class-limit	1	2	3	4	5	6	7	8
Frequency	95	295	350	150	40	30	20	20
Percentage cumulative frequency	9·5	40	75	90	95	97	99	100

These two distributions are negatively and positively skew, respectively.

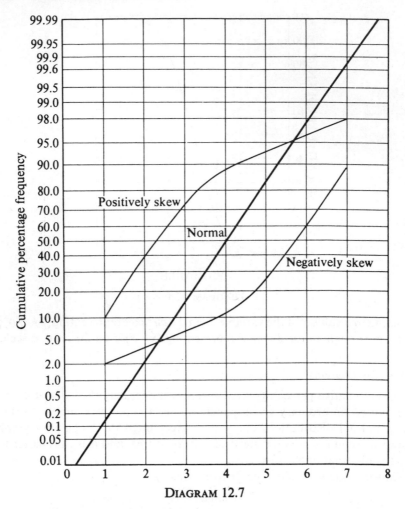

DIAGRAM 12.7

Plotting these on probability graph paper, we get two curves as in Diagram 12.7. Note that a positively skew distribution has the greater part of its convex side away from the *x*-axis, and a negatively skew distribution has its convex side towards the *x*-axis.

EXAMPLE

From the previous example (p. 159) on the depth of sapwood in telegraph poles, we have the following information.

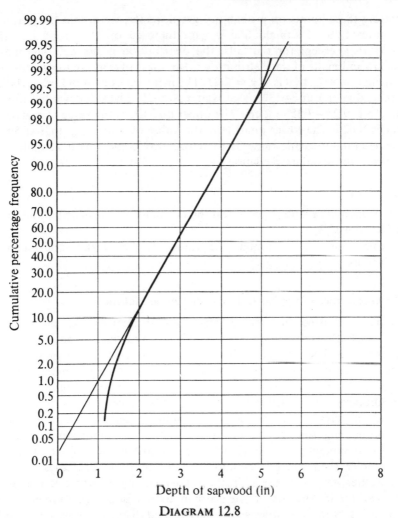

DIAGRAM 12.8

Upper class-limit	2·875	3·625	4·375	5·125	5·875	6·625
Cum. frequency	2	31	93	199	352	538
Percentage cum. freq.	0·15	2·26	6·79	14·53	25·69	39·27
	7·375	8·125	8·875	9·625	10·375	11·125
	731	919	1070	1193	1275	1323
	53·36	67·08	78·10	87·08	93·07	96·57
	11.875	12·625	13·375	14·125		
	1350	1364	1369	1370		
	98·54	99·56	99·93	100		

Plotting these points on probability graph paper, we obtain the graph in Diagram 12.8. A straight line is superimposed on the graph and this shows that, except for the tails, the distribution is normal. The mean, median and mode of the distribution is the value of x where the cumulative frequency is 50%, that is at $x = 7.25$. The mean plus one standard deviation is the value of x at 84.13%, that is at $x = 9.25$. The standard deviation is therefore, $9.25 - 7.25 = 2$ cm. The upper quartile is the value of x at 75%, that is 8.625. The lower quartile is the value of x at 25%, that is 5.875. The semi-interquartile range is therefore 1.375, and this is roughly two-thirds of the standard deviation.

Exercise 12.5

Use probability graph paper to test the normality of the distributions given in Exercise 12.3.

From your graphs estimate the values of their means and standard deviations.

The Occurrence of the Normal Distribution in Nature

We have discussed the normal distribution at great length, and it is only fair to ask if this distribution occurs often in practice. At one time it was thought that the normal distribution should be the most common type of distribution, but it is realised now that it is no more common than any other type. From experience it is found that the normal curve fits many distributions with sufficient accuracy to justify its use. The distributions are found in physics, nuclear sciences, biology, anthropology, sociology, education, engineering, etc., but probably its most important use is in the theory of sampling, where we will find the distribution of sample means approaches the normal distribution as the sample size increases.

Another important use is in the theory of errors, where it is used to estimate the importance of one error in comparison with another, or the estimating of the size of an error. The normal distribution was derived by Gauss as the law of the probability distribution of errors of measurement on the assumption that the true value of a measurement is the mean of a large number of observed values, that is that errors each side of the mean are equally likely. The 'law' depends on this assumption, and as no proof can be given of the assumption the usefulness of the 'law' can only be tested by experiment. It has been found that in particular cases the deviations from the mean follow the 'law' of errors very closely over the range mean plus or minus two and a half standard deviations, that is

98·76% of the observations, but generally shows discrepancies at the tails of the distribution. This is not surprising when we consider the number of observations at the tails of a normal distribution is small in comparison with the number of observations in the middle of the distribution.

Examining the distribution for the depth of sapwood in telegraph poles as drawn on the probability graph paper (Diagram 12.8), we see that the distribution obeys the normal law quite well over the range $(m-\sigma)$ to $(m+2\cdot5\sigma)$. As telegraph poles are selected poles we must not be surprised if the distribution does not follow the normal law at its extremes. Even so the range that obeys the normal law covers 83·5% of the observations.

EXAMPLE

Samples of size 300 are taken from an infinite population which is 5·5% defective. What is the probability of obtaining 20 or more defectives in a sample?

SOLUTION

This is a binomial distribution, but since the size of the sample is large we may treat it as a continuous normal distribution. The probability of obtaining a defective is 0·055. Therefore the probability of not obtaining a defective is $1-0\cdot055$, that is 0·945. Now $n = 300$. Therefore the mean number of defectives expected in a sample of size 300 is $np = 300 \times 0\cdot055 = 16\cdot5$, and the standard deviation is $\sqrt{(npq)} = \sqrt{(300 \times 0\cdot055 \times 0\cdot945)} = 3\cdot948$. Since the distribution is discrete the lower limit of the variate is 19·5.

Therefore, $\qquad X = (19\cdot5-16\cdot5)/3\cdot948 = 0\cdot760$

In the standardized normal curve the area to the right of $X = 0\cdot760$ is 0·7764. Therefore the probability of obtaining 20 or more defectives is

$$1-0\cdot7764 = 0\cdot2236$$

Exercise 12.6

1 A manufacturer of electric light bulbs which have a mean life of 1500 hours and a standard deviation of 250 hours wishes to guarantee his bulbs. If he is prepared to replace 5% of the bulbs sold, what should be the guaranteed life of the bulbs?

2 A machine produces rods of mean diameter 2·781 cm and standard

deviation 0·004 cm. Gauges reject all rods whose diameter is greater than 2·787 cm and less than 2·775 cm. What proportion of rods is rejected?

3   The following table gives the lengths of articles in cm, correct to the nearest cm. Calculate the theoretical frequencies for the same class-intervals for a normal curve with the same mean and standard deviation.

Height (cm)	60	61	62	63	64	65	66	67	68	69	70	71
Number	3	7	9	14	18	16	13	12	10	7	3	2

4   The masses of articles are normally distributed and 4·36% are under 30 kg and 6·3% are over 60 kg. Calculate the mean and standard deviation of the distribution.

5   In a normal distribution of mean 12 and standard deviation 3, what is the probability that the value of the variable (i) is greater than 15·3, (ii) is less than 5·7, (iii) lies between 8·7 and 14·1?

6   The following table gives the distribution of marks in a test. Calculate the theoretical frequencies for the same class-intervals for a normal curve with the same mean and standard deviation.

Class	1–10	11–20	21–30	31–40	41–50	51–60	61–70	71–80
Frequency	5	9	16	30	56	42	25	10

	81–90	91–100
	2	2

7   A lathe turns out rods of mean diameter 1·5 cm, and standard deviation 0·01 cm. Five per cent of the rods are rejected for over-size diameter. What is the critical upper diameter?

8   The mean strength of a certain material has been found to be 5000 N/cm^2 with a coefficient of variation of 20%. What strength can be guaranteed to be exceeded in 999 cases in 1000 samples?

9   In a normal distribution 10% of the items are under 25 units and 80% are under 73 units. Calculate the mean and standard deviation of the distribution.

10   Three hundred candidates sit a test of one hundred questions in which they are to answer 'Yes' or 'No'. If the pass mark is 45% and the questions are of equal value, how many candidates are expected to pass if all are guessing the answers?

11 If in the previous example there are three alternative answers to each question, how many candidates are expected to pass?

12 A normal distribution has a mean of 50 and a standard deviation of 4. Out of 500 items chosen at random, how many will be (i) between 51 and 52, (ii) between 44 and 45, (iii) greater than 48, (iv) between 54 and 60, (v) between 36 and 48, (vi) less than 56?

13 A coin is tossed 300 times. What is the probability of obtaining (i) 160 or more heads, (ii) less than 130 heads, (iii) between 140 and 160 heads, inclusive?

14 Packages rejected by a machine as underweight are weighed accurately and the records kept. They are as follows:

Weight of Package (N)	1·540–1·545	1·545–1·550	1·550–1·555	1·555–1·560
Frequency	2	5	13	23
	1·560–1·565	1·565–1·570		
	57	100		

Given these 200 rejects are from a sample of size 3200, estimate the probable percentage rejection if the machine is adjusted to make the rejection limit 1·5625 N.

15 The mean length of bolts made on a certain machine is 4 cm, with a standard deviation of 0·02 cm. Determine the critical limits of the length of the bolts if 2% are rejected for oversize and 2% for undersize.

16 Samples of size 200 are taken from an infinite normal population which is 5% defective. What is the probability of obtaining (i) 12 or more defectives, (ii) less than 5 defectives, (iii) from 6 to 15 defectives?

17 A large number of components produced by a machine have 3% rejected for oversize and 1% rejected for undersize when the upper and lower limits are 15·85 cm and 15·63 cm, respectively. Calculate the mean and standard deviation of the length of the components.

18 Six hundred investigators each ask two hundred persons how they will vote in an election. If the vote is equally divided between the two parties, X and Y, what is the number of investigators expected to report finding 120 or more X votes?

19 A product is sold in packets marked 1·5 kg. The average mass is

1·53 kg. If the distribution of masses is normal, what is the maximum value of the standard deviation if not more than one packet in two hundred is to be underweight?

20   The mean I.Q. of a large number of children of age 11 years is 100, and the standard deviation of the distribution is 16. What is the lower value of the range of the I.Q. of the top (i) 25%, (ii) 15%, (iii) 10%, (iv) 1% of the children?

What percentage of children have an I.Q. of 132 or more?

21   A machine is used to package sugar in 1 kg parcels. The standard deviation is 0·0025 kg. To what value should the machine be set so that at least 96% of the packages are over 1 kg in mass?

22   A parent population is normal and has a mean of 6 and a variance of 0·4. Calculate the probability of

(i) one observation exceeding 6·2,
(ii) each of three observations exceeding 6·2.

23   Find the mean and variance of the number, s, of successes in n trials, the probability of success at any trial being p.

A large cargo of lemons has, on the average, one bad lemon in five. Find an approximation to the probability that a random sample of 100 will contain 30 or more bad lemons.   (AEB)

24   Explain what is meant by the normal distribution and give examples of its occurrence.

Rods are made of nominal length of 4 cm but in fact they form a normal distribution with mean 4·01 cm and standard deviation 0·03 cm. Each rod costs sixpence to make and may be used immediately if its length lies between 3·98 and 4·02 cm. If its length is less than 3·98 cm, the rod is useless but has a scrap value of one penny. If its length exceeds 4·02 cm, it may be shortened and used at a further cost of twopence. Find the cost per usable rod.   (AEB)

25   Test the normality of the population in the table below using arithmetical probability graph paper. Explain your method.

Height of Boys in the Same Age Group

Height (metres)	1·5	1·525	1·55	1·575	1·6	1·625
Frequency	4	5	9	13	18	19
	1·65	1·675	1·7	1·725	1·75	1·775
	17	13	11	6	3	2

From your graph find

  (i) the arithmetic mean,
 (ii) the standard deviation,
(iii) the semi-interquartile range of the distribution.   (AEB)

26 The number of questions of equal difficulty attempted by a random sample of 1645 pupils in a two hour test are given in the following table:

No. of questions	0–	3–	6–	9–	12–	15–	18–	21–	24–	27–	30–
No. of pupils	3	22	75	194	331	402	321	192	79	22	4

Find (i) the median, (ii) the mode, and calculate (iii) the arithmetic mean, (iv) the standard deviation.

    It was decided in tests of equal difficulty about $2\cdot5\%$ of the candidates were to be given the opportunity of obtaining full marks. How many questions should be candidates be asked to answer in two hours?   (AEB)

27 The breaking stresses in kilonewtons per cm^2 of 100 specimens taken from a large consignment of steel bars are given in the following grouped frequency distribution:

Centre of interval	2·96	2·98	3·00	3·02	3·04	3·06
Frequency	6	14	30	28	15	7

Calculate the mean and standard deviation.

    Use the normal distribution with this mean and standard deviation to estimate the proportion of bars that will break at a stress less than 29·75 kilonewtons per cm^2.   (O and C)

28 State briefly what is meant by saying that a variable is normally distributed about a mean m with standard deviation s.

    Bolts are manufactured which are to fit in holes in steel plates. The diameter x of the bolts is normally distributed about 3·20 cm with standard deviation 0·03 cm; the diameter y of the holes is normally distributed about 3·31 cm with standard deviation 0·04 cm.

  (i) Find the proportion of bolts with diameter greater than 3·25 cm.
 (ii) Find the proportion of holes with diameter less than 3·25 cm.
(iii) If the bolts and holes are selected at random show that the proportion of bolts that will not fit the holes is $1\cdot39\%$.
(iv) Find the chance that a batch of 100 bolts will all fit their holes.

    (O and C)

29   The percentage marks obtained by 100 candidates in an examination are given in the following frequency table:

Centre of interval	4·5	14·5	24·5	34·5	44·5	54·5	64·5	74·5	84·5	94·5
Frequency	7	3	1	12	16	37	14	4	3	3

Calculate the mean and standard deviation.

Assuming that the marks obtained by 4000 candidates in this examination are normally distributed, estimate the number of candidates who will (a) obtain distinction (75 per cent or more), (b) fail (less than 35 per cent).

(O and C)

30   The breaking strengths in kilonewtons per cm^2 of 100 steel bars are given in the following frequency table:

Centre of interval	3·20	3·22	3·24	3·26	3·28	3·30	3·32	3·34	3·36	3·38
Frequency	1	4	11	13	21	24	12	9	3	2

Calculate the mean breaking strength and the standard deviation. Assuming that the population is normally distributed with this mean and standard deviation, calculate the percentage of bars that will break under loads between 3·25 and 3·32 kilonewtons per cm^2.   (O and C)

31   Rounds are fired from a machine gun at a fixed elevation and their ranges in metres beyond a mark 1400 metres from the gun are given in the following grouped frequency table:

Centre of interval (m)	60	80	100	120	140	160	Total
Frequency	3	19	32	37	19	10	120

Calculate the mean range and the standard deviation.

Assuming that the ranges are normally distributed with this mean and standard deviation, find the percentage of rounds that fall within 120 metres of the mark.   (O and C)

32   In a mechanism a plunger moves inside a cylinder and the difference between the diameters of the plunger and the cylinder must be at least 0·01 cm. Cylinders and plungers are manufactured separately. The diameters of the cylinders are normally distributed about a mean 2·025 cm with standard deviation 0·002; the diameters of the plungers are normally distributed about a mean 2·010 cm with standard deviation 0·003 cm.

(i) Find the proportion of cylinders that will have inadequate clearance for plungers of diameter 2·01 cm.

(ii) Find the proportion of plungers that will have inadequate clearance in cylinders of diameter 2·025 cm.
(iii) If plungers and cylinders are assembled at random, find the proportion of cases in which the clearance is insufficient.
(iv) If a set of ten plungers and cylinders are assembled in random pairs find the chance of the clearance being sufficient in all cases.

(O and C)

33 A train is scheduled to arrive at Paddington station at 10.26 a.m. each day. A passenger on 12 occasions recorded the number of minutes late as follows:

$$10 \quad 0 \quad 15 \quad 3 \quad -1 \quad 0 \quad 5 \quad 2 \quad 1 \quad 0 \quad -1 \quad 8$$

Calculate the mean arrival time of the train, the mean deviation of the times and the standard deviation.

Assuming that the times of arrival are normally distributed about the mean time with this standard deviation, estimate the chance of the train arriving on any one day, (a) more than 10 minutes late, (b) on or before the scheduled time. (O and C)

34 An athlete finds that in the high jump he can clear a height of 1·68 metres once in five attempts, and a height of 1·52 nine times out of ten attempts. Assuming the heights he can clear in various jumps form a normal distribution, estimate the mean and standard deviation of the distribution. Calculate the height the athlete can expect to clear once in one thousand attempts. (AEB)

35 An automatic machine produces bolts whose diameters are required to lie within the tolerance limits 1·26 cm to 1·28 cm. A random sample of bolts produced by the machine is found to have a mean diameter of 1·26 cm and a standard deviation of 0·005 cm. Assuming that the diameters are normally distributed, estimate the probability that any bolt produced by the machine will have a diameter outside the tolerance limits.

If the machine is adjusted to produce bolts of mean diameter 1·27 cm, the standard deviation being unaltered, estimate the percentage of bolts likely to be rejected on full inspection. (JMB)

Other Simple Distributions

There are many other continuous distributions in use and we will be meeting some later. Below are given three simple distributions and their constants.

If the student knows some simple calculus it is good practice for him to derive the constants for himself. If he has difficulty, the solutions are given in Mathematical Note 12.4.

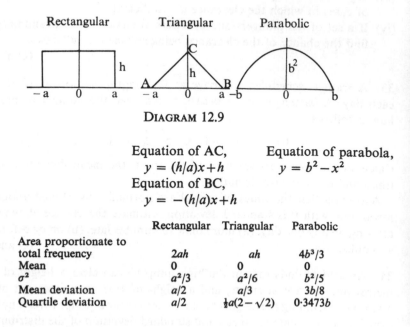

DIAGRAM 12.9

Equation of AC,   Equation of parabola,
$$y = (h/a)x + h$$ $$y = b^2 - x^2$$
Equation of BC,
$$y = -(h/a)x + h$$

	Rectangular	Triangular	Parabolic
Area proportionate to total frequency	$2ah$	ah	$4b^3/3$
Mean	0	0	0
σ^2	$a^2/3$	$a^2/6$	$b^2/5$
Mean deviation	$a/2$	$a/3$	$3b/8$
Quartile deviation	$a/2$	$\frac{1}{2}a(2-\sqrt{2})$	$0\cdot3473b$

Mathematical Note 12.1.   *The equation of the Normal Curve as a limiting form of the Binomial Distribution.*

The object is to replace the discrete binomial polygon by a continuous curve, such that the area between the two ordinates of the curve, that correspond to the ordinates of the boundaries of the corresponding interval of the frequency polygon, is equal to the frequency in that interval.

Suppose the ordinates of the binomial distribution are erected at equal horizontal intervals c, then on this scale the length required to represent σ will be $c\sqrt{npq}$. As n tends to infinity, c is to tend to zero in a way that keeps the length representing σ constant; that is, c is to vary inversely as $\sqrt{n}$. The range will be nc, and will vary as $\sqrt{n}$ since c varies inversely as $\sqrt{n}$, and will tend to infinity as n tends to infinity. Since we are interested in the symmetrical part of the distribution about the mean, it is more convenient to change the origin to the mean.

Let us suppose the value of the mean np is an integer. Since n is going to tend to infinity, the supposition does not impose any limitation on the work. Let y_r and y_{r+1} be the ordinates of the binomial distribution at r and $(r+1)$ intervals beyond the mean. Since the origin is at the mean np, y_r is the frequency of the $(np+r)$th number of successes. Therefore by the binomial distribution

$$y_r = {}_nC_{np+r} \times p^{np+r}q^{n-(np+r)}$$
$$= {}_nC_{np+r} \times p^{np+r}q^{nq-r}$$

and similarly,

$$y_{r+1} = {}_nC_{np+r+1} \times p^{np+r+1}q^{nq-r-1}$$

and

$$\frac{y_{r+1}}{y_r} = \frac{n-np-r-1+1}{np+r+1} \times \frac{p}{q}$$

$$= \frac{(nq-r)}{(np+r+1)} \times \frac{p}{q}$$

$$= \frac{(nq-r)p}{npq} \Big/ \frac{npq+(r+1)q}{npq}$$

Therefore, $\quad \log_e y_{r+1} - \log_e y_r = \log_e \frac{npq-rp}{npq} - \log_e \frac{npq+(r+1)q}{npq}$

$$= \log_e\left(1-\frac{r}{nq}\right) - \log_e\left(1+\frac{r+1}{np}\right)$$

As long as r is small compared with np and nq we can expand the right-hand side by means of the logarithmic series.

Therefore $\log_e y_{r+1} - \log_e y_r = -\dfrac{r}{nq} - \dfrac{r^2}{2n^2q^2} + \cdots$

$$-\frac{r+1}{np} + \frac{(r+1)^2}{2n^2p^2} - \cdots$$

$$= -\left[\frac{r}{nq} + \frac{r+1}{np}\right] + \left[\frac{(r+1)^2}{2n^2p^2} - \frac{r^2}{2n^2q^2}\right]$$

$$+ \text{ the further combinations of the terms in pairs}$$

$$= \frac{r(p+q)+q}{npq} + \frac{(r+1)^2q^2-r^2p^2}{2n^2p^2q^2} + \cdots$$

$$= -\frac{r+q}{npq} + \frac{(r+1)^2q^2-r^2p^2}{2n^2p^2q^2} + \cdots$$

$\log_e y_{r+1} - \log_e y_r$ may be written $\delta(\log_e y)$ or $\frac{1}{y}\delta y$, and c as δx. Therefore, dividing both sides of the equation by c, we have

$$\frac{1}{y}\frac{\delta y}{\delta x} = -\frac{r+q}{npqc} + \frac{(r+1)^2 q^2 - r^2 p^2}{2n^2 p^2 q^2 c} + \cdots$$

$$= \frac{(r+q)c}{npqc^2} + \frac{[(r+1)^2 q^2 - r^2 p^2]c^3}{2n^2 p^2 q^2 c^4} + \cdots$$

$npqc^2$ is a constant and is equal to the length representing σ^2. Therefore,

$$\frac{1}{y}\frac{\delta y}{\delta x} = -\frac{(r+q)c}{\sigma^2} + \frac{[(r+1)^2 q^2 - r^2 p^2]c^3}{2\sigma^4} + \cdots$$

Since $q < 1$, $(r+q)c$ is a value of x at a point between the ordinates y_r and y_{r+1} where x is measured from the mean, and as $n \to \infty$ and $c \to 0$ the limit of the right-hand side of the equation is $\frac{1}{y}\frac{dy}{dx}$ at the limiting position of the point $(r+q)c$. Therefore, in the limit we have

$$\frac{1}{y}\frac{dy}{dx} = -\frac{x}{\sigma^2}$$

the second and higher terms of the right-hand side tending to zero as $c \to 0$. Integrating the equation, we have

$$\log_e y = -\tfrac{1}{2}x^2 + \text{constant}$$

Therefore

$$y = y_0 \exp(-x^2/2\sigma^2)$$

and the constant y_0 is obviously the value of y when $x = 0$, i.e. the maximum ordinate.

Mathematical Note 12.2.

To solve the integral $I = \displaystyle\int_{-\infty}^{\infty} e^{-y^2} dy$

we can also write $I = \displaystyle\int_{-\infty}^{\infty} e^{-x^2} dx$

and multiplying $I^2 = \displaystyle\int_{-\infty}^{\infty} e^{-y^2} dy \int_{-\infty}^{\infty} e^{-x^2} dx$

$$= \int_{-\infty}^{\infty}\int_{-\infty}^{\infty} e^{-(x^2+y^2)} dx\, dy$$

This is a double integral taken over the complete area of a plane. Transferring the Cartesian coordinates to polar coordinates, $(x^2 + y^2)$ becomes r^2, and $dxdy$ is an element of area, dA, and in polar coordinates $dA = rd\theta dr$,

$$I^2 = \int\int e^{-r^2} rd\theta dr$$

and to cover the whole area of the plane the limits of θ must be 0 to 2π and r, 0 to ∞. Thus, we have

$$I^2 = \int_0^\infty e^{-r^2} rdr \times \int_0^{2\pi} d\theta$$

$$= 2\pi \int_0^\infty e^{-r^2} rdr$$

$$= \pi \int_0^\infty e^{-r^2} d(r^2)$$

$$= \pi \left[-e^{-r^2} \right]_0^\infty$$

$$= \pi$$

Therefore $\qquad I = \sqrt{\pi}$

If for y we substitute $\dfrac{x}{\sigma\sqrt{2}}$ then $dy = \dfrac{1}{\sigma\sqrt{2}} dx$. Therefore area under the normal curve

$$= y_0 \int_{-\infty}^\infty e^{-y^2} dy$$

$$= y_0 \sigma\sqrt{2} \int_{-\infty}^\infty \exp \frac{-x^2}{2\sigma^2} dx$$

$$= y_0 \sigma\sqrt{2}\sqrt{\pi}$$
$$= y_0 \sigma\sqrt{(2\pi)}$$

If the total frequency is N,

$$N = y_0 \sigma\sqrt{(2\pi)}$$

or $\qquad\qquad y_0 = \dfrac{N}{\sigma\sqrt{(2\pi)}}$

Mathematical Note 12.3.

The standard deviation of the distribution $y = y_0 \exp(-x^2/2\sigma^2)$ can be found by taking a value x of the variate and a small element dx, the frequency of the variate at x being $y\,dx$, and the mean square deviation from the mean equaling the variance. That is

$$\text{Variance} = \frac{1}{N}\int_{-\infty}^{\infty} x^2 y\,dx$$

$$= \frac{2}{N}\int_{0}^{\infty} x^2 y\,dx$$

$$= \frac{2}{N}y_0\int_{0}^{\infty} x^2 \exp\frac{-x^2}{2\sigma^2}\,dx$$

Let $t = \dfrac{x}{\sigma\sqrt{2}}$ then $dt = \dfrac{dx}{\sigma\sqrt{2}}$ and

$$\text{Variance} = \frac{2y_0}{N}\int_{0}^{\infty} 2\sigma^2 t^2 e^{-t^2}\,dt\,\sigma\sqrt{2}$$

$$= \frac{2\sqrt{2}y_0\sigma^3}{N}\int_{0}^{\infty} t(2te^{-t^2})\,dt$$

$$= -\frac{2\sqrt{2}y_0\sigma^3}{N}\int_{0}^{\infty} t\,\frac{de^{-t^2}}{dt}\,dt$$

$$= -\frac{2\sqrt{2}y_0\sigma^3}{N}\left(te^{-t^2} - \int_{0}^{\infty} e^{-t^2}\,dt\right)$$

$$= -\frac{2\sqrt{2}y_0\sigma^3}{N}\left(\left[te^{-t^2}\right]_{0}^{\infty} - \frac{\sqrt{\pi}}{2}\right)$$

$$= \frac{2\sqrt{2}y_0\sigma^3 \times \frac{1}{2}\sqrt{\pi}}{N}$$

But

$$y_0 = \frac{N}{\sigma\sqrt{(2\pi)}}$$

$$\text{Variance} = \frac{2\sqrt{2}\times\dfrac{N}{\sigma\sqrt{(2\pi)}}\sigma^3 \times \dfrac{\sqrt{\pi}}{2}}{N} = \sigma^2$$

Therefore the standard deviation $= \sigma$.

Mathematical Note 12.4.   *To calculate the mean deviation of the normal distribution*

The mean deviation

$$= \frac{1}{N} \int_{-\infty}^{\infty} |x| y \, dx$$

$$= \frac{2}{N} \int_{0}^{\infty} x y_0 \exp \frac{-x^2}{2\sigma^2} \, dx$$

$$= \frac{2y_0}{N} \int_{0}^{\infty} \sqrt{2}\sigma t e^{-t^2} \sqrt{2}\sigma \, dt$$

(where $t = \dfrac{x}{\sqrt{2}\sigma}$ as before) $\displaystyle = \frac{2y_0\sigma^2}{N} \int_{0}^{\infty} -e^{-t^2} d(-t^2)$

$$= \frac{-2y_0\sigma^2}{N} \left[e^{-t^2} \right]_{0}^{\infty}$$

$$= \frac{2y_0\sigma^2}{N}$$

But $\displaystyle y_0 = \frac{N}{\sigma\sqrt{(2\pi)}}$

Therefore mean deviation $\displaystyle = \frac{2N\sigma^2}{N\sigma\sqrt{(2\pi)}}$

$$= \sigma\sqrt{2}/\sqrt{\pi}$$

$$= 0 \cdot 7979\sigma$$

Therefore the mean deviation is nearly equal to $0 \cdot 8\sigma$, and this is the origin of the statement that the mean deviation is nearly equal to $4\sigma/5$.

Mathematical Note 12.5.   *The constants of three simple distributions.*

1 Rectangular Distribution

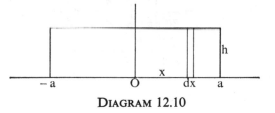

DIAGRAM 12.10

The total frequency is $2ah$. By symmetry the mean is at O. The frequency represented by the element dx distant x from O is hdx. The sum of the square of the deviations of these items is hx^2dx.

Therefore
$$\sigma^2 = \frac{1}{2ah}\int_{-a}^{a} hx^2dx = \frac{a^2}{3}$$

The mean deviation
$$= \int_{-a}^{a} \frac{h|x| \times dx}{2ah} = \frac{2h}{2ah}\int_{0}^{a} xdx = \frac{a}{2}$$

The quartile deviation. Let λ be the value of the quartile deviation. Then
$$\int_{0}^{\lambda} hdx = \tfrac{1}{4} \text{ of the area} = \tfrac{1}{2}ah$$

Therefore
$$h\lambda = \tfrac{1}{2}ah$$
$$\lambda = \tfrac{1}{2}a$$

2 The Triangular Distribution

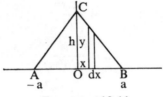

DIAGRAM 12.11

The equation of BC is
$$y = -\frac{h}{a}x + h$$

The equation of AC is
$$y = \frac{h}{a}x + h$$

Total frequency = area = ah. By symmetry the mean is at O.

$$\sigma^2 = \int_{0}^{a} \frac{x^2y}{ah}dx + \int_{-a}^{0} \frac{x^2y}{ah}dx$$

$$= \frac{2}{ah}\int_{0}^{a} x^2\left(-\frac{h}{a}x + h\right)dx$$

$$= \frac{2}{ah}\left[-\frac{h}{a}\frac{x^4}{4} + h\frac{x^3}{3}\right]_{0}^{a}$$

$$= \frac{a^2}{6}$$

The mean deviation $= 2\int_0^a \dfrac{xydx}{ah} = \dfrac{a}{3}$

The quartile deviation $= \int_0^\lambda ydx = \tfrac{1}{4}ah$

That is, $-\dfrac{h}{a}\dfrac{\lambda^2}{2}+h\lambda = \dfrac{ah}{4}$

or $\qquad\qquad 2\lambda^2 - 4a\lambda + a^2 = 0$

Therefore $\qquad\qquad \lambda = \tfrac{1}{2}a(2 - 2\sqrt{2})$
$\qquad\qquad\qquad\quad = 0\cdot2929a$

3 The Parabolic Distribution

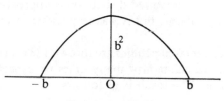

DIAGRAM 12.12

The equation of the parabola is $y = b^2 - x^2$

Total frequency $= \int_{-b}^{b} ydx = \dfrac{4b^2}{3}$

The variance $= \sigma^2 = \dfrac{3}{4b^2}\int_{-b}^{b} yx^2dx = \dfrac{b^2}{5}$

The mean deviation $= 2\times\dfrac{3}{4b^2}\int_0^b xydx = \dfrac{3b}{8}$

The quartile deviation

$$\int_0^\lambda ydx = \dfrac{b^3}{3}$$

Therefore $\qquad b^2\lambda - \dfrac{\lambda^3}{3} = \dfrac{b^3}{3}$

$$\lambda^3 - 3b^2\lambda + b^3 = 0$$
$$\lambda \simeq 0\cdot3473b$$

CHAPTER 13

Goodness of Fit and the χ^2 Test

In the chapters on the binomial, Poisson and normal distributions it was shown how to fit a theoretical distribution to an observed distribution. In the case of the Poisson and normal distributions it was possible to test graphically whether the observed distributions approximated to either of these standard distributions, but it was impossible to measure how good was the fit.

Let us assume six unbiased pennies are thrown 128 times. By the binomial distribution we can calculate how many times we expect 6 heads, 5 heads and 1 tail, and so on. The result is as follows:

6 heads	2
5 heads and 1 tail	12
4 heads and 2 tails	30
3 heads and 3 tails	40
2 heads and 4 tails	30
1 head and 5 tails	12
6 tails	2
Total	128

If we perform the experiment we should be very surprised if we obtained these exact frequencies, and would consider it quite reasonable if we obtained say, 5 heads and 1 tail 10, 11, 12, 13 or 14 times. The differences between the observed and theoretical frequencies are due to random sampling errors. If we perform the experiment a large number of times, we expect the average number of times 5 heads and 1 tail to appear to approach 12, but since the different possible frequencies of 5 heads and 1 tail are due to random sampling errors, we expect the frequencies to be normally distributed about 12 as a mean. Assuming the standard error or deviation about 12 to be, say, 2, then we can standardize any particular occurrence. Thus if the frequency is 15 the standardized variate is $\frac{1}{2}(15-12)$ which

186

equals $1\frac{1}{2}$, and if the frequency is 9 the standardized variate is $\frac{1}{2}(9-12)$ which equals $-1\frac{1}{2}$. If we do this for all the different occurrences of heads and ignore the signs of the standardized deviations from the mean we can obtain a standardized mean deviation which gives us an indication of the goodness of fit. But as in standard deviations it is found more convenient and mathematically advantageous to square the standardized deviations and then sum. The sum of the squares of the standardized deviations is denoted by χ^2 (a Greek letter called *chi*, pronounced kye). Therefore,

$$\chi^2 = \left(\frac{y_1 - \bar{y}_1}{\sigma_1}\right)^2 + \left(\frac{y_2 - \bar{y}_2}{\sigma_2}\right)^2 + \ldots \left(\frac{y_v - \bar{y}_v}{\sigma_v}\right)^2$$

where y_r is the observed value of the frequency of the rth variate, $\bar{y}_r$ the mean or expected value of the rth variate, σ_r the standard deviation of the observed frequencies for a series of observations of the rth variate, and v (Greek letter, *nu*), is the number of independent events or degrees of freedom for the observed distribution.

Degrees of Freedom

The number of degrees of freedom is the number of choices that can be made in fixing the values of the expected frequencies. Suppose we have a distribution with three events or classes. If we are asked to fix a frequency for each class and there are no restrictions, then we have to make three choices, say, f_1, f_2 and f_3. Thus we have three degrees of freedom, but if we are told that

$$f_1 + f_2 + f_3 = N$$

then we have one degree of freedom less, for if we fix f_1 and f_2, f_3 must equal $N - (f_1 + f_2)$.

When fitting a normal distribution to an observed distribution the two distributions must have the same mean, standard deviation and total frequency, that is there are three restrictions and the number of degrees of freedom is

$$v = \text{number of classes} - 3$$

If a Poisson distribution is fitted to an observed distribution the two must agree with respect to their means and total frequencies, and

$$v = \text{number of classes} - 2$$

In the case of a binomial distribution there are two possibilities.

(i) If the value of p is known then the two distributions must agree with respect to their frequencies. Thus, in a frequency distribution where we have to calculate the number of times, say, heads appear when six unbiased coins are tossed it is only necessary for us to know the total frequency to allow us to calculate the theoretical distribution. Therefore,

$$v = \text{number of classes} - 1$$

(ii) If the value of p is not known then the theoretical and observed distributions must agree with respect to their means and total frequencies, and this implies two restrictions. Therefore

$$v = \text{number of classes} - 2$$

The χ^2 Distribution

The distribution of χ^2 can be tabulated like the normal distribution, but the distribution varies appreciably with the number of degrees of freedom. To prevent the tables becoming too cumbersome the values of χ^2 for different values of v are only tabulated for selected values of probability. These probability levels are found sufficient for most problems. The shape of the χ^2 distribution is given in Diagram 13.1.

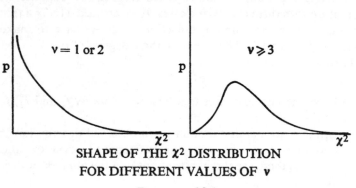

SHAPE OF THE χ^2 DISTRIBUTION
FOR DIFFERENT VALUES OF v

DIAGRAM 13.1

Simplification of χ^2

The calculation of χ^2 appears formidable, but fortunately it can be reduced to

$$\chi^2 = \Sigma \frac{(O-E)^2}{E} \qquad \text{(See Chapter 21)}$$

Where O is the observed value, (y_r in the previous formula), and E is the expected or mean value, ($\bar{y}_r$ in the previous formula).

Conditions for the Application of the χ^2 Test

Since we treat the different frequencies as if they belong to a normal or continuous distribution when they actually belong to a binomial or discrete distribution (a classified continuous distribution is similar to a discrete distribution), it necessitates the imposition of conditions before the χ^2 test can be applied. The conditions are:

(i) The total frequency of the distribution must be reasonably large—at least 50.
(ii) The theoretical frequency of each class should be at least 10. If frequencies fall below this number two or more classes should be combined.

EXAMPLE

Suppose we perform the experiment with the six pennies mentioned at the beginning of this chapter, and we obtain the following results:

Class	Frequency
6 heads	3
5 heads and 1 tail	7
4 heads and 2 tails	27
3 heads and 3 tails	55
2 heads and 4 tails	25
1 head and 5 tails	10
6 tails	1

Are we justified in saying the coins are unbiased?

SOLUTION

Tabulating the observed and expected frequencies together we have, assuming the coins are unbiased,

Class	6 h	5 h, 1 t	4 h, 2 t	3 h, 3 t	2 h, 4 t	1 h, 5 t	6 t
Observed frequency	3	7	27	55	25	10	1
Expected frequency	2	12	30	40	30	12	2

We notice the frequencies in the end classes are smaller than they should be for the application of the χ^2 test. Our first step is to combine these classes with their adjacent classes.

Class	5 or more h	4 h, 2 t	3 h, 3 t	2 h, 4 t	5 or more t
Observed frequency	10	27	55	25	11
Expected frequency	14	30	40	30	14

The conditions for the application of the test are now fulfilled and we proceed as follows:

Calculation of χ^2

O	E	$O-E$	$(O-E)^2$	$\dfrac{(O-E)^2}{E}$
10	14	-4	16	1·143
27	30	-3	9	0·300
55	40	15	225	5·625
25	30	-5	25	0·833
11	14	-3	9	0·643
				8·544

When estimating the expected frequencies we assumed $p = \frac{1}{2}$. Therefore the number of degrees of freedom is one less than the number of classes. The number of classes is now five. Therefore the number of degrees of freedom is

$$v = 5-1 = 4$$

Entering the χ^2 tables where $v = 4$ we see that $\chi^2 = 8·544$ lies approximately between the 5 and 10 per cent probability values, denoted by P. That is, the probability of getting a fit equal to or worse than the one we have considered would occur approximately 7·5 times in a hundred on the assumption that the coins are unbiased.

We must now make a decision. Is the probability of the observed event occurring so small that we may reject it as not due to random sampling errors, or is it possible for it to occur as often because of random sampling errors? The answer to this question will depend upon the type of problem under discussion. In most problems it is usual to assume that:

(i) if P is equal to, or greater than, 10%, the hypothesis is very probably true, because it is possible for the result to occur once in ten trials by chance, and this is not a rare event,

(ii) if P lies between 10% and 5% the hypothesis is probably true,

(iii) if P lies between the 5% and 2·5% levels the hypothesis is probably false,

(iv) if P lies between 2·5% and 1% the hypothesis is very probably false,
(v) if P is less than 1% the hypothesis is almost certainly false.

In the problem we have been discussing, P lies between 10% and 5% and the hypothesis is probably true, that is, our observed distribution is probably derived from unbiased coins. We must be careful not to argue from the converse that if the probability of the event happening is very large that we have proved the hypothesis. All we can say is that there is a strong likelihood of the hypothesis being correct. If the value of P is greater than 95% we should be suspicious of the result. Thus, if in the experiment just discussed we obtained an observed result equal to the expected result, we should consider it too good to be true, and should be suspicious of the experiment. In this case one of three possibilities may have happened:

(i) The data were carefully selected and not random.
(ii) The data were manufactured.
(iii) The hypothesis was manufactured to fit the data.

The Assessment of Results

The method employed in the above example to come to a decision sometimes disappoints students who expect a clear right or wrong from mathematics, but when we consider carefully, we find it is the method used every day of our lives to make judgments. If we wish to cross a road we look right, then left, then right again, judge the speed and distance of oncoming traffic and if we decide the probability of not being knocked down is large we decide it is safe to cross the road and proceed with confidence. Unfortunately we are sometimes wrong. The difference between the everyday application of this method and its use in statistics is that in everyday application we decide from subjective probability, but in statistics the probability is objective and numerically stated. For a further discussion on probability levels see page 234.

EXAMPLE

On page 123 there is a problem on the number of bulbs germinating in samples of five. The observed and expected number germinating were:

Number germinating	0	1	2	3	4	5
Observed frequency	1	6	14	33	31	15
Expected frequency	0	4	17	33	33	13

Do the observed results probably satisfy the theoretical binomial distribution?

SOLUTION

Rearranging to give the required magnitude to the frequencies, we have

Number germinating	2 or less	3	4	5
Observed frequency	21	33	31	15
Expected frequency	21	33	33	13
$O - E$	0	0	-2	2
$(O - E)^2$	0	0	4	4
$\dfrac{(O - E)^2}{E}$	0	0	0·121	0·3077

$$\chi^2 = \sum \frac{(O - E)^2}{E} = 0.429$$

To derive the theoretical binomial distribution we had to calculate the probability of the bulbs germinating, and since the total frequencies of the observed and expected distributions must be equal, there are two restrictions, and this makes the number of degrees of freedom equal to $4 - 2 = 2$. The value of χ^2 equal to 0·429 lies between the 10% and 95% value of P. Therefore we are justified in assuming the binomial gives a good fit to the distribution.

EXAMPLE

In 100 samples taken over a period the distribution of the number of defectives is given in the following table. Are we justified in assuming the number of defectives has remained constant?

Number of defectives per sample	0	1	2	3	4	5 or more
Number of samples	20	27	25	15	9	4

SOLUTION

The mean is 1·78 and the variance is 1·89. These are nearly equal so we try a Poisson distribution.

From the Poisson chart we get for $a = 1·8$,

> the probability of 1 or more successes is 0·83
> the probability of 2 or more successes is 0·54
> the probability of 3 or more successes is 0·26
> the probability of 4 or more successes is 0·11
> the probability of 5 or more successes is 0·035

Therefore,   the probability of obtaining 0 successes is 0·17
              the probability of obtaining 1 success is 0·29
              the probability of obtaining 2 successes is 0·28
              the probability of obtaining 3 successes is 0·15
              the probability of obtaining 4 successes is 0·075
              the probability of obtaining 5 or more successes is 0·035

Therefore, the expected values are:

Number of defectives	0	1	2	3	4	5 or more
Expected values	17	29	28	15	7·5	3·5

Rearranging to satisfy the conditions of the χ^2 test and proceeding with the calculation, we have,

Number of defectives	0	1	2	3	4 or more
Observed frequency	20	27	25	15	13
Expected frequency	17	29	28	15	11
$O - E$	3	−2	−3	0	2
$(O - E)^2$	9	4	3	0	4
$\dfrac{(O - E)^2}{E}$	0·529	0·138	0·321	0	0·364

$$\chi^2 = \sum \frac{(O - E)^2}{E} = 1·352$$

$$\nu = 5 - 2 = 3$$

From the tables we have $10\% < P < 95\%$. Therefore we are justified in assuming a good fit and that the proportion of the number of defectives has remained constant.

EXAMPLE

Adjusting the results of the fitting of a normal distribution to the depth of sapwood in telegraph poles (page 159) to satisfy the conditions for the application of the χ^2 test, we have,

Depth of sapwood	3·25 and less	4·0	4·75	5·5	6·25	7·0	7·75	8·5
Observed frequency	31	62	106	153	186	193	188	151
Expected frequency	46	53	92	138	179	202	199	170

	9·25	10·0	10·75	11·5	12·25 and over
	123	82	48	27	20
	127	82	46	22	16

Is the fitting of a normal distribution to the results satisfactory?

SOLUTION

$O-E$	-15	9	14	15	7	-9	-11
$(O-E)^2$	225	81	196	225	49	81	121
$\dfrac{(O-E)^2}{E}$	4·89	1·53	2·13	1·63	0·27	0·40	0·61
	-19	-4	0	2	5	4	
	361	16	0	4	25	16	
	2·12	0·13	0	0·09	1·14	1	

$$\chi^2 = \sum \frac{(O-E)^2}{E} = 15\cdot94$$

The distributions must agree with respect to their total frequencies, means and standard deviations. Therefore,

$$v = 13 - 3 = 10$$

From the χ^2 table we get $P = 10\%$. Therefore we are probably correct in assuming a good fit.

Exercise 13.1

1   Calculate the value of χ^2 for Examples 1 to 4 of Exercise 10.2.

2   Calculate the value of χ^2 for Examples 2 and 12 of Exercise 11.1.

3   Calculate the value of χ^2 for Examples 3 and 6 of Exercise 12.6.

4   In a large town, the number of road accidents reported daily over 300 working days gave the following results.

Number of accidents reported in a day (x)	0	1	2	3	4	5	6	7	8
Number of days (f)	17	43	69	68	50	28	13	8	4

Compare these frequencies with those of a Poisson distribution having the same mean and total frequency. Comment on the agreement.   (AEB)

CHAPTER 14

Sampling I

Introduction

Often it is impossible to examine every item of the parent population and so an investigator must make do with a sample. The essential characteristic of the items composing a sample is their random selection; that is, every item in the universe must have an equal chance of being selected. The problems of sampling are twofold: firstly, knowing the properties of the parent universe, what can be found out about the properties of a sample; and, secondly, knowing the properties of a sample, what can be found out about the parent population. The approach to these problems depends on the kind of parent population and the size of the sample. The parent population may be finite, infinite, or theoretical, and the sample small or large.

A finite population is one with a limited number of members, while an infinite population is one with an unlimited number of members, or, at least, so many that it can be considered infinite without introducing errors. A hypothetical population is one of which we can conceive without being able to attain in practice. Thus we can conceive of all the tosses of a coin without ever attaining them. The method of sampling has a bearing on the definition of the parent population. Thus if we draw a sample of 10 coloured balls from a bag containing, say, 500 balls and do not replace them, we are sampling from a finite population. But if the balls are replaced before drawing another sample, we can say we are sampling from an infinite population since any number of samples can be drawn without exhausting the parent population. What is a large or small sample will vary with the conditions of a problem, but a 'large' sample is generally one of 30 or more items when sampling for the mean, and generally 50 or more items in other cases. In many industrial applications if extreme accuracy is not important a large sample may be less than 30 items. The problem of sampling is estimating the reliability that can be placed on the value of any statistic. We must calculate the significance of that value.

195

Sampling Distributions

If we consider all the samples of size n that can be drawn from a parent population, it is possible to compute a sample statistic such as the mean or variance of each sample. The value of this sample statistic will vary from sample to sample, but we obtain a distribution which is known as the sampling distribution of the statistic. Frequently these distributions are normally, or near normally, distributed and it is possible to apply the theory of normal distributions to them. At another time they are differently distributed and it is necessary to compute a different set of tables to determine the probability of the value of the statistic under discussion. We had an example of this when discussing χ^2.

Sampling Distribution of the Means. Finite Population

The most common statistics are the arithmetic mean and standard deviation. The relationships between the mean and standard deviation of the parent population and the mean and standard deviation of the sample means are as follows:

Let m, and μ, denote the means and s, and σ, the standard deviations of a sample and the parent population, respectively, and let the mean and standard deviation of the distribution of sample means be μ_m and σ_m. Let the samples be of size n and the parent population of size N. Then if sampling is without replacement

$$\mu_m = \mu, \quad \text{and} \quad \sigma_m = \frac{\sigma}{\sqrt{n}}\sqrt{\left(\frac{N-n}{N-1}\right)}$$

(see Mathematical Note 14.1)

It can be proved that if the parent population is normally distributed, the means of the samples will be normally distributed. Even if the parent population is not normally distributed, the distribution of the means of the samples still tends to be normally distributed if the size of the samples and the parent population are sufficiently large—that is, $n \geqslant 30$ and $N > 2n$. In the case of small samples the distribution of the means is not normal. The above relationships give us information about the means of samples if the distribution of the parent population is known.

EXAMPLE

A population consists of the four digits 1, 2, 3, 6 and all possible samples of size two are drawn from it without replacement. Calculate,

(i) the mean of the parent population,
(ii) the standard deviation of the parent population,
(iii) the mean of the means of the samples,
(iv) the standard deviation of the means of the samples.

SOLUTION

(i) The mean of the parent population.

$$\mu = \frac{1+2+3+6}{4} = 3$$

(ii) The standard deviation of the parent population.

$$\sigma^2 = \frac{(1-3)^2+(2-3)^2+(3-3)^2+(6-3)^2}{4}$$

$$= 3 \cdot 5$$
$$\sigma = 1 \cdot 87$$

(iii) The mean of the means of the samples.
There are $_4C_2 = 6$ samples of size two that can be drawn without replacement. They are

$$(1, 2); \quad (1, 3); \quad (1, 6); \quad (2, 3); \quad (2, 6); \quad (3, 6)$$

We cannot draw (1, 2) and (2, 1) and so on, for these are the same sample if there is no replacement. The means of the samples are:

$$1 \cdot 5; \quad 2; \quad 3 \cdot 5; \quad 2 \cdot 5; \quad 4; \quad 4 \cdot 5;$$

and the mean of these sample means is

$$\mu_m = 18 \cdot 0/6 = 3 = \mu$$

(iv) The variance of the sample means.

$$\sigma_m{}^2 = \frac{(1 \cdot 5-3)^2+(2-3)^2+(3 \cdot 5-3)^2+(2 \cdot 5-3)^2+(4-3)^2+(4 \cdot 5-3)^2}{6}$$

$$= \frac{7}{6}$$

Substituting in

$$\frac{\sigma^2}{n}\left(\frac{N-n}{N-1}\right)$$

we have

$$\frac{3 \cdot 5}{2}\left(\frac{4-2}{4-1}\right) = \frac{7}{6} = \sigma_m{}^2$$

Therefore we have verified the relationships.

EXAMPLE

A machine makes 500 similar components with a mean weight of 4·03 N and a standard deviation of 0·20 N. Calculate the probability that a random sample of 60 components from this group will have a mean weight of (a) between 4·0 and 4·1 N (b) more than 4·05 N. Sampling is without replacement and the weights of the components are normally distributed.

SOLUTION

The mean of the samples = $\mu_m = \mu = 4\cdot03$ N. The standard deviation of the mean of the sample means equals

$$\sigma_m = \frac{\sigma}{\sqrt{n}}\sqrt{\left(\frac{N-n}{N-1}\right)}$$
$$= \frac{0\cdot20}{\sqrt{60}}\sqrt{\left(\frac{500-60}{500-1}\right)}$$
$$= 0\cdot0242$$

(a) Since the weights of the components are normally distributed the means of the samples will be normally distributed. Standardizing the mean weights 4·0 and 4·1 N we have,

$$X = \frac{m-\mu_m}{\sigma_m} = \frac{4\cdot0-4\cdot03}{0\cdot0242} = -1\cdot2397$$

and
$$\frac{4\cdot1-4\cdot03}{0\cdot0242} = 2\cdot8926$$

The required probability equals the area under the normal curve between the ordinates at $X = -1\cdot2397$ and $X = 2\cdot8926$, that is 0·8906.

(b) Standardizing the weight 4·05 N, we have,

$$X = \frac{4\cdot05-4\cdot03}{0\cdot0242} = 0\cdot8264$$

The required probability equals the area under the normal curve to the right of the ordinate at $X = 0\cdot8264$, that is 0·2043.

Infinite Parent Populations

Since $\mu_m = \mu$ is independent of N, μ_m remains unchanged if N becomes infinite, but

$$\sigma_m = \frac{\sigma}{\sqrt{n}}\sqrt{\left(\frac{N-n}{N-1}\right)}$$

$$= \frac{\sigma}{\sqrt{n}}\sqrt{\left(\frac{1-n/N}{1-1/N}\right)} \quad \text{(dividing numerator and denominator by } N)$$

$$= \frac{\sigma}{\sqrt{n}} \text{ as } n \text{ tends to infinity}$$

EXAMPLE

An infinite population consists of equal numbers of the four digits 1, 2, 3, 6 and samples of size two are drawn from this population. Calculate,

(i) the mean of the population,
(ii) the standard deviation of the population,
(iii) the mean of the means of the samples,
(iv) the standard deviation of the means of the samples.

SOLUTION

(i) There is an equal number of each digit, so the mean of the population will equal the mean of the digits; that is,

$$\mu = \tfrac{1}{4}(1+2+3+6) = 3$$

(ii) Similarly the standard deviation of the population will be equal to the standard deviation of the digits. Therefore,

$$\sigma^2 = 3\cdot5$$

(iii) There are sixteen different samples that can be drawn from the parent population. They are,

$$
\begin{array}{cccc}
(1, 1) & (1, 2) & (1, 3) & (1, 6) \\
(2, 1) & (2, 2) & (2, 3) & (2, 6) \\
(3, 1) & (3, 2) & (3, 3) & (3, 6) \\
(6, 1) & (6, 2) & (6, 3) & (6, 6)
\end{array}
$$

This time it is possible to select (1, 2) and (2, 1) and so on because there is an infinite number of each digit and if a large number of samples is selected there will be near enough an equal number of each sample. The mean of the sample means will therefore be the mean of the sixteen samples listed above. The sample means are,

$$\begin{array}{cccc} 1 & 1\cdot5 & 2 & 3\cdot5 \\ 1\cdot5 & 2 & 2\cdot5 & 4 \\ 2 & 2\cdot5 & 3 & 4\cdot5 \\ 3\cdot5 & 4 & 4\cdot5 & 6 \end{array}$$

Therefore, $\mu_m = \dfrac{\text{sum of sample means}}{16} = \dfrac{48}{16} = 3 = \mu$

(iv) Similarly the variance of the distribution of the sample means will equal the variance of the sixteen sample means.

$$\sigma_m{}^2 = \frac{(1-3)^2+(1\cdot5-3)^2+(2-3)^2 \text{ and so on}}{16}$$

$$= 1\cdot75$$

Now, $\sigma^2 = 3\cdot5$

Therefore $\dfrac{\sigma^2}{n} = \dfrac{3\cdot5}{2} = 1\cdot75 = \sigma_m{}^2$

Therefore the relationships are verified.

EXAMPLE

The standard deviation of a very large population is 2·33 cm. If random samples of 100 items are drawn from this population, what is the probability that a sample mean will differ from the true mean by 0·66 cm or more?

SOLUTION

Since the population can be taken as infinite, the standard deviation of the sample means will be

$$\frac{\sigma}{\sqrt{n}} = 2\cdot33/10 = 0\cdot233 \text{ cm}$$

Since the size of the sample is large, the means of the samples will be normally distributed. Standardizing the difference between the sample mean and the true mean, we have since $\mu_m = \mu$

$$\frac{(m-\mu)}{\sigma_m} = \frac{0\cdot66}{0\cdot233} = 2\cdot8326$$

Therefore, the probability of obtaining a sample mean that is greater than the true mean by 0·66 or more, is from the table of the normal distribution, 0·00238. Similarly the probability of obtaining a sample mean that is less than the true mean by 0·66 or more is 0·00238. Therefore the probability of obtaining a sample mean that differs from the true mean, that is, is greater than or less than 0·66, is

$$0·00238 + 0·00238 = 0·00476$$

Therefore, it is very unlikely that we obtain a sample mean with this difference.

EXAMPLE

In the previous example what must be the minimum difference between the sample mean and the population mean if the chances of obtaining this difference is to be 99·6%?

SOLUTION

The chances of obtaining a positive difference is to be $\frac{1}{2}(100 - 99·6)$, or, 0·2%. From the table for the normal distribution function the value of the standardized variable corresponding to the probability 0·2% or 0·002 is 2·88. From the previous example the standard deviation of the sample means is 0·233. Therefore the minimum difference is

$$\pm 0·233 \times 2·88 = \pm 0·671 \text{ cm}$$

Exercise 14.1

1 A normal distribution has a mean of 30 and a standard deviation of 3. Calculate the probability that if 16 items are drawn at random their mean (a) is 32 or more, (b) lies between 28 and 31·5.

2 The heights of 4000 plants are normally distributed with a mean of 68·2 cm and standard deviation 2·67 cm. If 50 samples consisting of 36 plants each are obtained without replacement, in how many samples is the mean expected to (a) lie between 67·8 and 68·9 cm, (b) be less than 67·5 cm?

3 The weights of 1600 components are normally distributed with a mean of 23·26 N and a standard deviation of 0·056 N. If a large number of random samples of size 49 are drawn from this population, determine the expected mean and standard deviation of the sampling distribution of means if sampling is done (a) with replacement, (b) without replacement.

4 Solve question 3 if the population consists of 160 components.

5   The masses of packages of a commodity have a mean of 30 kg and a standard deviation of 0·4 kg. Calculate the probability that the mass of a carton of 64 packages selected at random will exceed 1928 kg.

6   The length of a component is specified as 20 cm with a standard deviation of 0·45 cm. A purchaser of these components takes a sample of 100 and finds its mean to be 19·85 cm. Is the purchaser justified in saying the rods are not to specification?

7   The lengths of the forearms of a large number of adult males were measured and found to have a mean of 47·9 cm with a standard deviation of 1·6 cm. Suppose random samples of n adult males are taken and the mean and standard deviation of the sample means are found, how large is n to be for the standard deviation of the sample means to be less than 0·25 cm?

8   A parent population is normal and has a mean of 6 and a variance of 0·6. What is the probability of (a) the average of 4 observations exceeding 6·2, (b) the average of 25 observations exceeding 6·2?

9   In the previous question what must be the size of the random sample if the probability that the mean of the sample will not be greater than the population mean by more than 0·2 is 95%?

10   A sample of 400 items has a mean of 18·0. What is the chance that this sample mean comes from a population whose mean is 18·9 and standard deviation 5·8?

Distribution of the Sum or Difference of Two Independent Variates

Suppose we have two *independent* populations the n_1 variates composing the first population being denoted by x, and the n_2 variates of the second population by y; that is

$$x = x_1, x_2, x_3, \ldots x_{n_1}$$
$$y = y_1, y_2, y_3, \ldots y_{n_2}$$

Let us form a new population whose variates, **X**, are all the possible variates that can be formed by the addition of one variate from each population x and y. That is,

$$\mathbf{X} = x_1+y_1, x_1+y_2, x_1+y_3, \ldots x_1+y_{n_2}+$$
$$x_2+y_1, x_2+y_2, x_2+y_3, \ldots x_2+y_{n_2}+$$

$$\vdots \qquad \vdots \qquad \vdots \qquad \vdots$$

$$x_{n_1}+y_1, x_{n_1} \ y_2, x_{n_1}+y_3, \ldots x_{n_1}+y_{n_2}$$

If μ_X, σ_X, μ_x, σ_x and μ_y, σ_y are the means and standard deviations of the populations $\mathbf{X}$, x and y, respectively, then

$$\mu_X = \mu_x + \mu_y$$

and
$$\sigma_X = \sqrt{(\sigma_x{}^2 + \sigma_y{}^2)}$$

(See Mathematical Note 14.2)

Let us check these two relationships by means of a numerical example.

Let
$$x = 1, 2, 3 \quad \text{and} \quad y = 3, 4$$

Then
$$\mu_x = 2, \qquad \sigma_x = \sqrt{(2/3)}, \qquad \mu_y = 3\tfrac{1}{2}, \qquad \sigma_y = \sqrt{\tfrac{1}{4}}$$
$$\mathbf{X} = 1+3+ \qquad 1+4+ \qquad 2+3+ \qquad 2+4+ \qquad 3+3+ \qquad 3+4$$

and

$$\mu_X = 5\tfrac{1}{2}; \qquad \sigma_X = \sqrt{(11/12)}$$
$$\mu_x + \mu_y = 2 + 3\tfrac{1}{2} = 5\tfrac{1}{2} = \mu_X$$
$$\sqrt{(\sigma_x{}^2 + \sigma_y{}^2)} = \sqrt{(\tfrac{2}{3} + \tfrac{1}{4})} = \sqrt{(11/12)} = \sigma_X$$

In a similar way we can illustrate that if a new population $\mathbf{Y}$ is formed from all the possible values of $(x-y)$, then

$$\mu_Y = \mu_x - \mu_y \quad \text{and} \quad \sigma_Y = \sqrt{(\sigma_x{}^2 + \sigma_y{}^2)}$$

(Note well that $\sigma_X = \sigma_Y$)
The numerical illustration of these two relationships is similar to the previous illustration.

It is important to note that if two populations are normally distributed then the distribution of the sum or difference of the variates is also normally distributed. A proof of this can be found in statistics books dealing with the mathematical theory of statistics.

The relationships hold for more than two populations. Generally if

$$ax = ax_1, ax_2, \ldots ax_{n_1}$$
$$by = by_1, by_2, \ldots by_{n_2}$$
$$cz = cz_1, cz_2, \ldots cz_{n_z}$$

and so on and a new population $\mathbf{X}$ is formed where

$$\mathbf{X} = \pm ax \pm by \pm cz \pm \ldots$$

then
$$\mu_X = \pm a\mu_x \pm b\mu_y \pm c\mu_z \pm \ldots$$

and
$$\sigma_X{}^2 = a^2\sigma_x{}^2 + b^2\sigma_y{}^2 + c^2\sigma_z{}^2 + \ldots$$

(See Mathematical Note 14.3)

EXAMPLE

Four components are assembled linearly. The length of each component is distributed normally with a mean and standard deviation of $1{\cdot}00$, $0{\cdot}02$; $1{\cdot}50$, $0{\cdot}03$; $0{\cdot}75$, $0{\cdot}01$; and $3{\cdot}60$, $0{\cdot}04$ cm, respectively. Between what limits do 99% of the components lie?

SOLUTION

The mean length of the assembled components

$$= 1{\cdot}00 + 1{\cdot}50 + 0{\cdot}75 + 3{\cdot}60 = 6{\cdot}85 \text{ cm}$$

The variance of the length of the assembled components

$$= (0{\cdot}02)^2 + (0{\cdot}03)^2 + (0{\cdot}01)^2 + (0{\cdot}04)^2 = 0{\cdot}0030$$

In the standardized normal curve 99% of the area is included between the normal variates $\pm 2{\cdot}58$. Therefore the required limits are

$$6{\cdot}85 \pm 2{\cdot}58 \sqrt{(0{\cdot}0030)} = 6{\cdot}85 \pm 0{\cdot}14 \text{ cm}$$

Sampling Means

The populations x and y can be sample means from two infinite populations with means and deviations μ_1, σ_1, and μ_2, σ_2, respectively. If the sizes of the samples from the first and second populations are n_1 and n_2, respectively, and μ_{m_1}, μ_{m_2}, σ_{m_1}, σ_{m_2}, the means and standard deviations of the sample means from the first and second populations, respectively, then

$$\mu_{m_1} = \mu_1; \qquad \mu_{m_2} = \mu_2$$

and

$$\sigma_{m_1} = \sigma_1/\sqrt{n_1}; \qquad \sigma_{m_2} = \sigma_2/\sqrt{n_2}$$

Also

$$\mu_{m_1 \pm m_2} = \mu_{m_1} \pm \mu_{m_2} = \mu_1 \pm \mu_2$$

and

$$\sigma_{m_1 \pm m_2} = \sqrt{(\sigma_{m_1}{}^2 + \sigma_{m_2}{}^2)} = \sqrt{\left(\frac{\sigma_1{}^2}{n_1} + \frac{\sigma_2{}^2}{n_2}\right)}$$

The results also hold if the sampling is from finite populations with replacement. If the populations are finite and there is no sample replacement, then

$$\sigma_{m_1} = \frac{\mu_1}{\sqrt{n_1}} \sqrt{\left(\frac{N_1 - n_1}{N_1 - 1}\right)} \quad \text{and} \quad \sigma_{m_2} = \frac{\mu_2}{\sqrt{n_2}} \sqrt{\left(\frac{N_2 - n_2}{N_2 - 1}\right)}$$

where N_1 and N_2 are the number of items in the populations.

EXAMPLE

Two manufacturers supply the same component to a factory. The factory has been testing the tensile strength of samples from each manufacturer for a long period and has found the standard deviations of the tensile strengths of the components to be 96·9 N per cm² from the first manufacturer and 126·3 N per cm² from the second manufacturer. The factory finds 36 components from the first factory have a mean tensile strength of 2933 and 48 components from the second factory have a mean tensile strength of 2885. Calculate whether the mean tensile strength of the components is the same from both manufacturers.

SOLUTION

With the usual notation,

$$m_1 = 2933; \quad m_2 = 2885; \quad \sigma_1 = 96\cdot9;$$
$$\sigma_2 = 126\cdot3; \quad n_1 = 36; \quad n_2 = 48$$

The standard deviation of the difference of the means

$$= \sigma_{m_1-m_2} = \sqrt{\left[\frac{\sigma_1{}^2}{n_1} + \frac{\sigma_2{}^2}{n_2}\right]}$$

$$= \sqrt{\left[\frac{(96\cdot9)^2}{36} + \frac{(126\cdot3)^2}{48}\right]}$$

$$= 24\cdot56$$

If there is no difference between the means of the two populations, $m_1 - m_2$ will have zero mean and standard deviation 24·56 and will be normally distributed. The standardized variate is

$$\frac{(m_1 - m_2) - 0}{\sigma_{m_1-m_2}} = \frac{2933 - 2885}{24\cdot56} = 1\cdot954$$

From the table of the normal distribution the area to the right of the ordinate at 1·954 is 0·0254. Since the question asks if the tensile strength of the components is the same from both manufacturers and not whether the first is greater than the second, the problem requires what is called a two-tailed test. Therefore the probability is 2 × 0·0254 or 0·0508. This result could occur about once in twenty trials and is not uncommon enough to be conclusive. Hence the evidence suggests there may be no difference between the tensile strengths of the components but it is not overwhelmingly conclusive and more tests are desirable.

One and Two-tailed Tests

The previous example is described as a two-tailed test because the question asks if the tensile strength of the components is the same from both manufacturers. That is, we wish to know what is the probability that the tensile strength of the components from the first manufacturer is greater than of those from the second, or of those from the second manufacturer greater than of those from the first, allowing for random sampling errors. So the total probability is the sum of the probabilities of either event happening. If we were only interested in whether the tensile strengths of the components from the first manufacturer are greater than those from the second we would only need a one-tailed test. Further examples of one and two-tailed tests will be met with in the text.

EXAMPLE

Two firms A and B manufacture similar components with a mean breaking strength of 3000 N and 2500 N and standard deviations of 200 N and 100 N, respectively. If random samples of 100 components of manufacturer A and 50 components of manufacturer B are tested, what is the probability that the components from manufacturer A will have a mean breaking strength which is at least (a) 450, (b) 575 N more than the components of manufacturer B?

SOLUTION

Let m_A and m_B denote the mean breaking strength of samples A and B, respectively Then

$$\mu_{m_A - m_B} = \mu_{m_A} - \mu_{m_B} = 3000 - 2500 = 500 \text{ N}$$

and

$$\sigma_{m_A - m_B} = \sqrt{\left[\frac{\sigma_A^2}{n_A} + \frac{\sigma_B^2}{n_B}\right]}$$

$$= \sqrt{\left[\frac{(200)^2}{100} + \frac{(100)^2}{50}\right]}$$

$$= 24 \cdot 49$$

The standardized variate for the difference in means is

$$\frac{(m_A - m_B) - (\mu_{m_A - m_B})}{\sigma_{m_A - m_B}} = \frac{(m_A - m_B) - 500}{24 \cdot 49}$$

and will be normally distributed.

(a) If $m_A - m_B = 450$, the standardized variate $= (450 - 500)/24.49 = -2.042$. The required probability is the area under the normal curve to the right of the normal variate -2.042, that is 0.9794.

(b) If $m_A - m_B = 575$, the standardized normal variate $= (575 - 500)/24.49 = 3.062$. The required probability is the area under the normal curve to the right of the ordinate at 3.062, that is,

$$1 - 0.9988 = 0.0012$$

EXAMPLE

If A and B both toss 50 unbiased coins what is the chance of A tossing 10 or more heads more than B?

SOLUTION

The probability of tossing a head is $\frac{1}{2}$. The mean number of heads A or B expects is $np = 50 \times \frac{1}{2} = 25$, with a standard deviation $\sqrt{(npq)} = \sqrt{(50/4)}$. Therefore, the standard deviation of the difference of the means

$$\sigma_{m_A - m_B} = \sqrt{(\tfrac{50}{4} + \tfrac{50}{4})} = 5$$

Since the number of items in each distribution is 50 and $p = q$, we can use the normal curve to calculate the probabilities of the discrete distribution without any appreciable error. We must remember that on a continuous variable basis 10 or more heads must be taken as 9.5 or more heads. Therefore, since the difference in the expected means is zero the standardized variate equals $(9.5 - 0)/5 = 1.9$, and the required probability is the area under the normal curve to the right of the ordinate at 1.9, that is,

$$1 - 0.9713 = 0.0287$$

(Note this is a one-tailed test because the question asks what is the chance of A tossing 10 or more heads more than B? It would be a two-tailed test if the question asked what is the chance of the difference in the number of heads between A and B being 10 or more?)

EXAMPLE

A normal distribution A has a mean of 45 cm and a standard deviation of 2 cm, while a normal distribution B has a mean of 44 cm and a standard deviation of 1.5 cm. What is the probability that

(i) two variables from A differ by 1·5 cm or more,
(ii) two variables from B differ by 1·5 cm or more,
(iii) one variable from A and one from B differ by 1·5 cm or more?

SOLUTION

(i) A may be considered as two distributions x and y each with mean 45 cm and standard deviation 2 cm.
What we wish to find is the probability that $x - y$ is 1·5 or more. The distribution $x - y$ has a mean 0 and a variance of $2^2 + 2^2$ i.e. 8. Therefore its standard deviation is $\sqrt{8}$, i.e. $2\sqrt{2}$. The standardized variate for $x - y$ = 1·5 is

$$(1·5 - 0)/2\sqrt{2} = 0·5303$$

The probability of a standardized variate for $x - y = 0·5303$ is 0·702. Therefore the probability of $x - y \geqslant 1·5$ is $1 - 0·702 = 0·298$. Since the same probability will apply for $y - x \geqslant 1·5$, the total probability that two variables from A differ by 1·5 cm or more is $2 \times 0·298 = 0·596$.

(ii) Similarly if B is considered as two distributions x and y the mean of $x - y$ is 0 and its standard deviation is $\sqrt{(2 \times 2·25)} = 2·121$. Therefore, the standardized variate for $x - y = 1·5$ is

$$\frac{1·5 - 0}{2·121} = 0·7072$$

The probability of a standardized variate equal to 0·7072 is 0·7602. Therefore the probability of $x - y \geqslant 1·5$ is $1 - 0·7602 = 0·2398$. Therefore the probability that two variables from B differ by 1·5 cm is $2 \times 0·2398 = 0·5796$.

(iii) Let a variate from the distribution A be denoted by x and one from B by y. The mean and standard deviation of the x's are 45 and 2 cm and of the y's 44 and 1·5 cm, respectively. If we form a distribution $x - y$ its mean is 1 and its standard deviation $\sqrt{(4 + 2·25)} = 2·5$. Therefore the standardized variate for $x - y = 1·5$ is

$$\frac{1·5 - 1}{2·5} = 0·2$$

The probability of obtaining a standardized variate $\geqslant 0·2$ is $1 - 0·5793 = 0·4207$. If we form a distribution $y - x$ its mean and standard deviation is -1 and 2·5, respectively. The standardized variate for $y - x = 1·5$ is

$$\frac{1 \cdot 5 - (-1)}{2 \cdot 5} = 1$$

Therefore the probability of obtaining a variate from $y - x \geqslant 1 \cdot 5$ is $1 - 0 \cdot 8413 = 0 \cdot 1587$. Therefore the total probability of obtaining a variable from the distributions A and B differing by $1 \cdot 5$ is $0 \cdot 4207 + 0 \cdot 1587 = 0 \cdot 5794$.

Exercise 14.2

1 In a large national examination the marks were normally distributed with a mean of 45 and a standard deviation of 16. Two batches of sample scripts were selected consisting of 52 and 72 scripts. What is the probability their mean scores differ by (a) 2 or more marks, (b) 10 or more marks, (c) between 3 and 5 marks?

2 The probability of a person voting for a certain candidate in an election is 7/10. What is the probability if two canvassers interviewed random samples of voters of size 300, that their results would indicate a difference of more than 5% in the proportions voting for the candidate?

3 In a box there are 65 red balls and 35 white balls, all similar except for colour. Two samples of 40 balls are selected at random, the first sample being replaced before drawing the second. What is the probability that the samples differ by 10 or more red balls?

4 The mean lengths of three components are $2 \cdot 48$, $3 \cdot 60$ and $6 \cdot 23$ cm with standard deviations of $0 \cdot 21$, $0 \cdot 43$, $0 \cdot 52$ cm, respectively. If the three components are joined linearly, find (a) the mean and (b) the standard deviation of the compound article.

5 Car batteries have a mean voltage of $12 \cdot 0$ volts with a standard deviation of $0 \cdot 2$ volts. What is the probability that six such batteries will have a combined voltage of 71 volts?

6 Two works are supposed to be producing bricks with the same transverse strength. Samples are tested from the two works with the following results.

	Works 1	Works 2
Number in random sample	332	286
Mean strength in N per cm^2	998	1024
Standard deviation in N per cm^2	236	204

Are the bricks from Works 1 weaker than those from Works 2?

7   A bar is made up of two parts assembled linearly. The mean lengths of the parts are 3·4 and 4·2 cm, and the standard deviations 0·012 and 0·014 cm respectively. Calculate the mean length and the standard deviation of the assembled bars.

8   Three components A, B and C of mean weights 50, 10 and 5 N, and standard deviations of 2, 0·5 and 0·3 N, respectively, are assembled to form an article of mean weight 150 N. There are 2 components of A, 3 of B and 4 of C. If the finished article weighs less than 144 N or more than 155 N, it is rejected. Calculate the percentage rejected if the weight of each component is normally distributed.

9   Four 400 metre runners have mean times of 50·4, 50·8, 51·2 and 51·8 seconds with standard deviations of 0·8, 0·9, 1·1 and 1·2 seconds, respectively. Assuming their individual times are independent of each other, are normally distributed and that their time for a 4×400 metre relay is the sum of their individual times for 400 metres, calculate the probability of their breaking a relay record of 3 min 20·2 s.

10   If x and y are independent random variables, show that the mean of $x - y$ is the difference of the means of x and y and that the variance of $x - y$ is the sum of the variances of x and y.
    A sample of 500 taken at random from a normal population is found to have a mean 6 and a standard deviation 2·31. Another sample of 800 taken at random from a second normal distribution has a mean 5 and standard deviation 3·08. Estimate the probability that, if a random selection of one value is taken from each population, the value from the second population will be greater than that from the first.   (AEB)

11   Show that the variance of the mean of a random sample of n values taken independently from an infinite population of variance σ^2 is σ^2/n. Metal rods are delivered in large batches to a customer. When production is under control, the rods have a mean length of 2·132 cm, with a standard deviation of 0·001 cm. A random sample of 90 from a batch is found to have a mean length of 2·1316 cm. Report on the likelihood that the production is still under control.   (AEB)

12   In a manufacturing process on the average one item in seven is defective. In a random sample of 25 what is the mean number defective? What is the most likely number of defectives?

In fourteen successive samples of 25 a total of 60 defective items was found. Discuss the likelihood that the fraction defective in the production is still one seventh. (AEB)

13 The heights of men in Ruritania form a normal distribution with mean 168·75 cm and standard deviation 6·5 cm. The heights of the women of Ruritania are normally distributed with mean 162·5 cm and standard deviation 6 cm. What is the probability that, if a Ruritanian man and woman are taken at random, the woman is taller than the man? (AEB)

14 A group of boys and a group of girls were given a test in mathematics. The arithmetic mean scores, standard deviations, and numbers in the groups were as follows:

	Boys	Girls
Arithmetic mean score	121·3	118·2
Standard deviation	12·1	10·2
Number in group	84	61

Is the difference between the arithmetic mean scores of the boys and girls significant? Give reasons for your conclusions. (AEB)

15 A boy keeps a daily record of the times it takes him to walk to school in the morning. Over a period of some months his average time is 17·2 minutes with a standard deviation of 2·14 minutes. He then decides to record only a weekly average for a five-day school week. Assuming all other conditions remain unchanged what should he expect the new average and standard deviation to be when based on the weekly records? (AEB)

16 The tensile strength of a large number of metal die-castings has an arithmetic mean value of 29·2 and a standard deviation of 1·3, the units being 100 N/cm². If random samples of size 100 are taken from this population and the arithmetic mean tensile strength of each sample taken, between what limits would 95% of the means of the samples be expected to lie? (AEB)

17 A machine is intended to make insulators 3 cm long. These are satisfactory provided their standard deviation is not more than 0·01 cm. As the machine gets older its accuracy is tested by finding the mean lengths of sample batches of 20 insulators. It is found that the standard deviation of the means of a number of batches is 0·005 cm. State, with reasons, what conclusion you draw about the condition of the machine. (C)

18   A tractor drawbar is intended to have a breaking strain of 120 kN. It is known that, because of small variations in the casting, the breaking strain varies between individual bars with standard deviation 4 kN force. A sample of 5 bars gave a mean breaking strain of 115 kN, is there significant evidence (at the 5% level) of a change in breaking strain?

Obtain a limit which the new mean breaking strain may be said with 95% confidence to exceed.   (C)

19   The heights of boys in a given age group are normally distributed with mean 150 cm. and standard deviation 5 cm; the heights of girls in the same age group are normal with mean 147·5 cm and standard deviation 4·5 cm. Determine as accurately as you can the probabilities of differences in height greater than 2 cm between:

(a) two boys in the group,
(b) two girls in the group,
(c) a boy and a girl in the group.   (C)

20   A variable is observed 10 times and the mean and standard deviation of these observations are 4 and 1·2; it is then observed 20 times more, these observations having mean and standard deviation 3·8 and 1·4. Why are the two pairs of values not the same? This being the case, what is the use of observing the variable at all? Calculate the standard deviation of the mean in the two cases.

How would you calculate a combined mean for the two sets of observations? Why not use $\frac{1}{2}(4+3·8)$?   (C)

21   The breaking strain of a certain type of chain is expected to be 1·300 kN and the standard deviation is known to be 0·040 kN. The mean breaking strain of 9 test lengths was 1·323 kN; was there significant evidence of a change in breaking strain, and if so at what level of significance?

The manufacturing process was slightly modified, the standard deviation remaining unchanged: the first 9 test lengths gave breaking strains 1·254, 1·301, 1·344, 1·318, 1·394, 1·335, 1·262, 1·337, 1·362 kN. Is there significant evidence that the breaking strain has been increased, and if so at what level of significance?   (C)

22   Explain the use of the formula $\sqrt{(pq/n)}$ for testing the significance of a proportion.

A coin was tossed 20 000 times. In the first 10 000 tosses 5075 heads were obtained. If the experiment had stopped there, would this have been significant evidence of bias?

In the second test of 10 000 tosses, 5100 heads were obtained. To what conclusion does this lead? Give reasons for your answers. (C)

23 If hens of a certain breed lay eggs on four days a week, on the average, find how many days during a season of 200 days a poultry keeper with eight hens of this breed will expect to receive at least six eggs.

If in fact he received at least six eggs on 42 days during the season of 200 days, would this evidence suggest that any factor other than chance was operating? (C)

24 The standard deviation of the number of articles produced in an hour by a factory worker is 14 for all workers, but the mean number produced varies from worker to worker. The numbers produced by a new employee in a representative hour on each of ten successive days were: 108, 124, 92, 113, 129, 146, 117, 103, 131, 128. Do these values show that this worker's rate of production is, at the 5% level of significance, below the factory average of 127 articles per hour?

Determine the two factory average rates from which the average of these ten observations would be judged just to differ at the 5% level of significance. How many more observations would be needed to reduce to 14 the difference between the two rates defined in this way? (C)

25 In an investigation of preference between pre-packed and fresh cut bacon, 69 housewives preferred pre-packed and 52 preferred fresh cut bacon. Does this result provide evidence (at the 5% level of significance) of a difference in acceptability?

Find those values of the percentage of housewives preferring pre-packed which could be regarded as reasonable, that is which do not differ from the observed proportion at the 5% level of significance. (The value of a standardized normal deviate which is exceeded in 5% trials is 1·96.) (C)

26 Explain briefly what is meant by the standard error of the mean.

In a chemical plant the acid content of the effluent from the factory is measured frequently. From 400 measurements the acid content in grammes per 100 litres of effluent is recorded in the following frequency distribution:

Acid content	12	13	14	15	16	Total
Frequency	5	52	235	74	34	400

(i) Find the mean acid content and the standard error of the mean.

(ii) Assuming a normal distribution of the acid content, give 95 per cent confidence limits for the mean acid content of the effluent.

(iii) Is the result consistent with a mean acid content of 14·13 grammes per 100 litres obtained from tests over several years?   (O and C)

27   Steel bars are ordered with the following specifications:

(a) each bar to have a breaking strength of at least 2·95 kN per cm²,
(b) the mean breaking strength to be not less than 3·0 kN per cm².

A random sample of 50 bars from a large consignment is tested and found to have a mean breaking strength of 2·995 kN per cm², with a standard deviation 0·021 kN per cm². Assuming that the distribution of breaking strengths is normal, find

(i) what proportion of the consignment would be expected to have a breaking strength below specification,

(ii) the values of 95 per cent confidence limits for the mean breaking strength of the consignment,

(iii) the chance of obtaining a mean as low as 2·995 kN per cm² if the consignment meets the specification (b) exactly.   (O and C)

28   In an examination there are 4000 candidates. In Paper I the marks may be taken as a continuous and normal distribution with mean 47·5 and standard deviation 12·5.

(a) Estimate the numbers of candidates who will (i) obtain less than 40 marks, (ii) obtain marks between 36 and 44.

(b) Estimate the mark which will be exceeded by 10 per cent of the candidates.

(c) If a particular examiner marks 100 scripts taken at random and awards marks whose mean is 49·5, is there evidence that his marking is more lenient than that of the other examiners?

(d) In Paper II the mean mark is 50·5 and the standard deviation 15 and the total mark for the examination is the average for the two papers. How many candidates will have a total mark less than 40?   (O and C)

29   Explain briefly what is meant by the standard error of the mean. Steel rods are being manufactured in quantity to a specified length of

6·20 cm and tolerance limits laid down for the production are 6·15 cm to 6·25 cm. The lengths of the rods in a random sample of 100 are measured to the nearest 0·01 cm and their dimensions in hundredths of a cm above 6·14 cm are given in the following frequency table:

Length	0	1	2	3	4	5	6	7	Total
Frequency	3	10	25	30	20	9	2	1	100

(i) Find the mean of the sample and the standard error of this mean.

(ii) Assuming a normal distribution, show that the mean length of rod produced will be between 6·165 and 6·173 cm approximately.

(iii) Show that about 8 per cent of the rods produced will fail to meet the tolerance.   (O and C)

30   Find the variance of the mean of a random sample of n values drawn independently from a large population having a variance σ^2.

One hundred men of the same race are taken at random, and their heights in centimetres are found to have a mean 175·3 and variance 45·2. Find 95% and 98% confidence limits for the mean height of the men of the race.   (AEB)

31   One thousand men are weighed, and the mass in each case is given to the nearest half kilogramme. The rounding error (true mass *less* recorded mass) is x half-kilogrammes. What would you expect to be the distribution of the random variable x?

Find the mean and variance of x.

If y half-kilogrammes is the true mass of a man, his mass is recorded as $z = (y-x)$ half-kilogrammes where x, y may be regarded as statistically independent variables. Show that the variance of z is the sum of the variances of x and y, and show how the variance of y can be estimated from that of z.   (AEB)

32   It is found that over a certain period at one telephone exchange 200 subscribers taken at random made a total of 13 248 calls. During the same time, a random sample of 300 subscribers at another exchange made a total of 20 922 calls. The standard deviation of the number of calls made by a subscriber at either exchange in the period is 8. Is there any evidence of a difference between the subscribers at the two exchanges in their average frequency of calls?

Find 95% confidence limits for the mean number of calls made in the period per subscriber at each exchange.   (AEB)

33   Throughout Great Britain during 1958 the question 'At about what age did you start to smoke a cigarette (or a pipe) a day for as long as a year?' was put to a random sample of 100 male smokers. The following table was drawn up from the answers received:

Age at which the man started to smoke	Number of men
15	27
16	19
17	12
18	16
19	5
20–24	15
25–29	4
30–34	1
35–39	1

Using 17 years as an arbitrary origin, or otherwise, calculate the mean age and the standard deviation of the ages at which the men began to smoke.

Given that the corresponding mean and the standard deviation of 100 female smokers are 23·8 years and 9·8 years, respectively, determine whether or not the difference between the means is significant.   (JMB)

34   During a test on a steam turbine, two observers A and B, using separate manometers a and b, respectively, took two sets of pressure readings. Both instruments were connected to the same orifice, and, at equal intervals of time, observations were made alternately by each observer. The following table summarizes the results:

	Observer A using manometer a	Observer B using manometer b
Number of readings	100	100
Mean of readings (cm of mercury)	35·71	36·37
Standard deviation (cm of mercury)	2·03	2·03

(i) If it could be assumed that each set of 100 readings is normally distributed and that each set could be regarded as a random sample of the variations in pressure, determine whether or not the difference between the means is significant at the 5 per cent level.

(ii) Assuming that the 100 readings made by observer A approximated very closely to a normal distribution, construct their frequency distribution using the following class-intervals: less than 33·0, 33·0–, 34·3–, 35·6–, 36·9, 38·2–, 39·5 and over.

(iii) It was found that the readings taken by observer A agreed closely with those expected of a normal distribution but that those taken by B

did not. Assuming that both observers were equally reliable, state any deductions that might be made from the results of the test. If the main object of the test is to register variations in the pressure turbine, suggest ways of improving the test when it is repeated. (JMB)

Mathematical Note 14.1. *The Mean and Variance of Sample Means.*

Before considering the general case let us consider the simpler problem of a population of four values x_1, x_2, x_3 and x_4 and samples of size three. Suppose the mean of x_1, x_2, x_3 and x_4 is zero and their standard deviation is σ. Taking the mean as zero simplifies the problem and makes no difference to the problem since it only supposes that a constant quantity has been subtracted from each value of the variate.

Therefore,

$$\tfrac{1}{4}(x_1+x_2+x_3+x_4) = 0$$

or

$$\sum_1^4 x_r = 0$$

Let x_1, x_2, x_3 be a random sample of size three.
The number of samples that can be selected is 4, [i.e. $_4C_3$]. They are

$$x_1, x_2, x_3; \qquad x_1, x_2, x_4; \qquad x_1, x_3, x_4; \qquad x_2, x_3, x_4$$

Let m_1, m_2, m_3 and m_4 be the means of these samples and μ_m the mean of the four sample means m_r.

Then

$$x_1+x_2+x_3 = 3m_1; \qquad x_1+x_2+x_4 = 3m_2;$$

$$x_1+x_3+x_4 = 3m_3; \qquad x_2+x_3+x_4 = 3m_4;$$

and

$$m_1+m_2+m_3+m_4 = 4\mu_m$$

$$\sum_1^4 m_r = 4\mu_m \qquad [\text{i.e. } _4C_3\mu_m] \tag{1}$$

We see from the samples that if we choose any particular value of x there are three ways [i.e. $_{4-1}C_{3-1}$] in which the other two values of x can be chosen to make up a sample of size three. Therefore, each value of x occurs three [i.e. $_{4-1}C_{3-1}$] times in the samples, and therefore the sum of all the x's in the sample is

$$3(x_1+x_2+x_3+x_4) = 3\sum_1^4 x_r = {}_{4-1}C_{3-1} \times \sum_1^4 x_r$$

The sum of all the x's occurring in the samples is also given by

$$3m_1 + 3m_2 + 3m_3 + 3m_4 \quad \text{or} \quad 3\sum_1^4 m_r$$

Therefore $\qquad 3\sum_1^4 m_r = 3\sum_1^4 x_r = 0 \quad \left(\text{Since } \sum_1^4 x_r = 0\right)$

Therefore $\qquad \sum_1^4 m_r = 0$

and therefore, $\qquad \mu_m = 0 \qquad\qquad$ [by (1)]

Therefore the mean of the sample means is equal to the mean of the population.

Since the mean of the parent population is zero, x_1, x_2, x_3, x_4 are also the deviations of the items from the mean. Therefore, if σ is the standard deviation of the parent population

$$\sigma^2 = \tfrac{1}{4}(x_1^2 + x_2^2 + x_3^2 + x_4^2)$$

or $\qquad\qquad 4 \times \sigma^2 = \sum_1^4 x_r^2 \qquad\qquad\qquad\qquad (2)$

Similarly, since the mean of the means of the samples is also zero and there are 4 [i.e. $_4C_3$] samples whose deviations from their mean are m_1, m_2, m_3, m_4 the variance of the sample means is

$$\sigma_m^2 = \frac{1}{4}\sum_1^4 m_r^2$$

or $\qquad\qquad 4\sigma_m^2 = \sum_1^4 m_r^2$

$$= [m_1^2 + m_2^2 + m_3^2 + m_4^2]$$

$$= \frac{1}{3^2}\sum(x_1 + x_2 + x_3)^2 \text{ over all the samples}$$

$$(x_1 + x_2 + x_3)^2 = x_1^2 + x_2^2 + x_3^2 + 2x_1x_2 + 2x_1x_3 + 2x_2x_3$$
$$(x_1 + x_2 + x_4)^2 = x_1^2 + x_2^2 + x_4^2 + 2x_1x_2 + 2x_1x_4 + 2x_2x_4$$
$$(x_1 + x_3 + x_4)^2 = x_1^2 + x_3^2 + x_4^2 + 2x_1x_3 + 2x_1x_4 + 2x_3x_4$$
$$(x_2 + x_3 + x_4)^2 = x_2^2 + x_3^2 + x_4^2 + 2x_2x_3 + 2x_2x_4 + 2x_3x_4$$

Therefore, on adding, each x^2 occurs 3 [i.e. $_{4-1}C_{3-1}$] times and each product $2x_jx_k$ ($j \neq k$) occurs twice [i.e. $_{4-2}C_{3-2}$]

Therefore
$$4\sigma_m^2 = \frac{1}{3^2}\left[3\sum_1^4 x_r^2 + 2\times 2\sum_1^4 x_j x_k\right]$$

i.e. $\frac{1}{3}[_{(4-1)}C_{(3-1)}\Sigma x_r^2 + 2\times_{4-2}C_{3-2}x_j x_k]$ (3)

Also $(x_1+x_2+x_3+x_4)^2 = x_1^2+x_2^2+x_3^2+x_4^2$
$$+2x_1x_2+2x_1x_3+2x_2x_3+2x_2x_4+2x_3x_4$$

or
$$\left[\sum_1^4 x_r\right]^2 = \sum_1^4 x_r^2 + 2\sum_1^4 x_j x_k \qquad \text{where } j\neq k$$

But
$$\sum_1^4 x_r = 0$$

Therefore, substituting in (3)
$$4\sigma_m^2 = \frac{1}{3^2}\left[_{4-1}C_{3-1}\sum_1^4 x_r^2 - _{4-2}C_{3-2}\sum_1^4 x_r^2\right]$$

$$= \frac{1}{3^2}[_{4-1}C_{3-1} - _{4-2}C_{3-2}]\sum_1^4 x_r^2$$

$$= \frac{1}{3^2}[_{4-1}C_{3-1} - _{4-2}C_{3-2}]4\sigma^2 \qquad \text{[by (2)]} \quad (4)$$

Now $_4C_3 - (4/3)_3C_2$ and $_3C_2 = (3/2)_2C_1$

Therefore (4) becomes

$$4\sigma_m^2[\text{i.e. } _4C_3\sigma_m^2] = \frac{1}{3^2}\,_3C_2[1-\tfrac{2}{3}]4\sigma^2$$

or
$$\sigma_m^2 = \tfrac{1}{3}[1-\tfrac{2}{3}]\sigma^2$$

$$= \frac{\sigma^2}{3}\left[\frac{4-3}{4-1}\right]$$

or
$$\frac{\sigma^2}{3}\left[\frac{N-n}{N-1}\right]$$

It is now a simple matter to write out the general proof. Suppose we have a parent population of N values, $x_1, x_2, x_3 \ldots x_N$ whose mean is zero and standard deviation σ.

Then
$$\sum_1^N x_r = 0$$

Let $x_1, x_2 \ldots x_n$ be a random sample of size n. The number of random samples of size n that can be selected is $_NC_n$. Let m_r be the mean of a sample and μ_m the mean of the $_NC_n$ sample means. Then

$$n \times m_r = \sum_1^n x_r \quad \text{and} \quad _NC_n\mu_m = \Sigma m_y$$

If we choose any value of x there are $_{N-1}C_{n-1}$ ways in which the other $n-1$ values of x can be chosen to make up a sample of size n. Therefore each value of x occurs $_{N-1}C_{n-1}$ times, and the sum of all the x's in all the samples is given by $_{N-1}C_{n-1} \sum_1^N x_r$ and also by $n\Sigma m_r$.

Therefore,
$$n\Sigma m_r = {_{N-1}C_{n-1}} \sum_1^N x$$

$$= 0 \quad \left[\text{Since } \sum_1^N x_r = 0\right]$$

Therefore $\Sigma m_r = 0$

and therefore $\mu_m = 0$

Thus the mean of the sample means is equal to the mean of the population.

$$\sigma^2 = \frac{1}{N} \sum_1^N x_r^2 \quad \left[\text{Since } \left(\frac{\Sigma x}{N}\right)^2 = 0\right]$$

or $$N \times \sigma^2 = \sum_1^N x_r^2$$

Since the mean of the sample means is also zero and there are $_NC_n$ samples the variance of the sample means is

$$\sigma_m^2 = \frac{\Sigma m_r^2}{_NC_n} \tag{5}$$

that is, $_NC_n\sigma_m^2 = \Sigma m_r^2$

$$= \frac{1}{n^2} \Sigma(x_1 + x_2 + \ldots x_n)^2 \qquad \text{over all the samples}$$

$$= \frac{1}{n^2} \left[_{N-1}C_{n-1}\Sigma x_r^2 + 2_{N-2}C_{n-2}x_jx_k\right] \qquad (j \neq k)$$

Since the square of each x occurs in $_{N-1}C_{n-1}$ samples and the product $2x_jx_k$ occurs in $_{N-2}C_{n-2}$ samples

$$\left(\sum_1^N x_r\right)^2 = \sum_1^N x_r^2 + 2\sum_1^N x_j x_k$$

But

$$\sum_1^N x_r = 0$$

Therefore

$$2\sum_1^N x_j x_k = -\sum_1^N x_r^2$$

Therefore

$$_NC_n\sigma_m^2 = \frac{1}{n^2}\left[_{N-1}C_{n-1}\sum_1^N x_r^2 - _{N-2}C_{n-2}\sum_1^N x_r^2\right]$$

$$= \frac{1}{n^2}\left[_{N-1}C_{n-1} - _{N-2}C_{n-2}\right]\sum_1^N x_r^2$$

$$= \frac{1}{n^2}\left[_{N-1}C_{n-1} - _{N-2}C_{n-2}\right]N\sigma^2$$

Now $\quad _NC_n = \dfrac{N}{n}\times _{N-1}C_{n-1}\quad$ and $\quad _{N-1}C_{n-1} = \dfrac{N-1}{n-1}\times _{N-2}C_{n-2}$

Therefore

$$\sigma_m^2 = \frac{1}{n}\left[1 - \frac{n-1}{N-1}\right]\sigma^2$$

$$-\frac{\sigma^2}{n}\left[\frac{N-n}{N-1}\right] \tag{6}$$

or

$$\sigma_m = \frac{\sigma}{\sqrt{n}}\sqrt{\left(\frac{N-n}{N-1}\right)}$$

Mathematical Note 14.2.   The Mean and Variance of the Sum or Difference of Two Populations.

Suppose we have two independent distributions ax and by, (a and b are constants) of N_1 and N_2 variates, respectively. Let their means and standard deviations be μ_x, σ_x and μ_y, σ_y. Let us now form a new population, X, whose variates are the sum of all possible pairs of variates that can be formed by taking one variate from each of the populations ax and by.

$$\begin{aligned}
X = \; & ax_1 + by_1, \quad ax_1 + by_2, \quad \ldots \quad ax_1 + by_{N_2} \\
& ax_2 + by_1, \quad ax_2 + by_2, \quad \ldots \quad ax_2 + by_{N_2} \\
& \;\;\vdots \qquad\qquad\quad \vdots \qquad\qquad\qquad \vdots \\
& ax_{N_1} + by_1, \quad ax_N + by_2, \quad \ldots \quad ax_{N_1} + by_{N_2}
\end{aligned}$$

The mean, μ_x, of the new population is formed by summing all these items and dividing by the number of items $N_1 \times N_2$. That is

$$\mu_X = \frac{1}{N_1 \times N_2} [N_2 \Sigma ax + N_1 \Sigma by]$$

$$= a \frac{\Sigma x}{N_1} + b \frac{\Sigma y}{N_2}$$

$$= a\mu_x + b\mu_y$$

Since the mean of a distribution can be made zero by subtracting a constant amount equal to the mean from each item, and this will not affect the standard deviation of the distribution, we can assume that the two distributions x and y have zero means.

Therefore **X** will have a zero mean.

$$\Sigma X^2 = (ax_1 + by_1)^2 + (ax_1 + by_2)^2 + \ldots (ax_1 + by_{N_2})^2$$
$$\vdots \qquad\qquad \vdots \qquad\qquad\qquad \vdots$$
$$+ (ax_{N_1} + by_1)^2 + (ax_{N_1} + by_2)^2 + \ldots (ax_{N_1} + by_{N_2})^2$$
$$= N_2(a^2x_1{}^2 + a^2x_2{}^2 + \ldots a^2x_{N_1})^2 + N_1(b^2y_1{}^2 + b^2y_2{}^2 + \ldots b^2y_{N_2}{}^2)^2$$
$$+ 2ab(x_1y_1 + x_1y_2 + \ldots x_1y_{N_2}$$
$$\vdots \qquad\qquad \vdots$$
$$x_{N_1}y_1 + x_{N_1}y_2 + \ldots x_{N_1}y_{N_2})$$
$$= a^2 N_2 \sum_1^{N_1} x^2 + b^2 N_1 \sum_1^{N_2} y^2 + 2ab \sum_1^{N_1} x \times \sum_1^{N_2} y \qquad (1)$$

Since **X**, x and y have zero means, Σx and Σy equal zero. Therefore

$$\sigma_X{}^2 = \frac{\Sigma X^2}{N_1 \times N_2}$$

$$= a^2 \frac{\Sigma x_2{}^2}{N_1} + b^2 \frac{\Sigma y^2}{N_2} \quad \text{by (1)}$$

$$= a^2\sigma_x{}^2 + b^2\sigma_y{}^2$$

This theorem can be extended to three or more populations, and to the sum or difference of the populations. Thus

Mean of $(\pm ax \pm by \pm cz \ldots) = \pm a\mu_x \pm b\mu_y \pm c\mu_z \pm \ldots$

and the

Variance of $(\pm ax \pm by \pm cz \pm \ldots) = a^2\sigma_x{}^2 + b^2\sigma_y{}^2 + c^2\sigma_2{}^2 + \ldots$

Appendix: The Variance of the Population $X = f(x \times y)$

If X is not a linear function of x and y there is a method of finding the variance of X for small errors in x and y. No proof is given, but the relationship is

$$\sigma_X^2 = \left(\frac{\partial f}{\partial x}\right)_m^2 \sigma_x^2 + \left(\frac{\partial f}{\partial y}\right)_m^2 \sigma_y^2$$

where $(\partial f/\partial x)_m$ and $(\partial f/\partial y)_m$ stand for the values of the partial differential coefficients $\partial f/\partial x$ and $\partial f/\partial y$ at the mean values of x and y.

EXAMPLE

The area of the curved surface of a cone is given by $A = \pi r l$ where r is the radius of the base and l the slant height. r and l have mean values of 5 and 9 cm with standard deviations of 0·05 and 0·06 cm, respectively. Calculate the mean and variance of the curved surface of the cone.

SOLUTION

$$\text{Mean area} = \pi \times 5 \times 9 = 45\pi \text{ cm}^2$$

$$\frac{\partial A}{\partial r} = \pi l \qquad \frac{\partial A}{\partial l} = \pi r$$

$$\sigma_A^2 = \left(\frac{\partial A}{\partial r}\right)_m^2 \sigma_r^2 + \left(\frac{\partial A}{\partial l}\right)_m^2 \sigma_l^2$$

$$= \pi^2 l^2 (0·05)^2 + \pi^2 r^2 (0·06)^2$$

$$= 0·11025\pi^2 \quad (l = 9; r = 5)$$

$$\sigma_A = 0·332\pi$$

Exercise 14.3

1   If the mean values of l, h, b and r are 3, 2, 2·5 and 1·5 cm, and their standard deviations 0·05, 0·02, 0·03 and 0·01 cm, respectively, calculate the mean value and standard deviation of A if

(a) $A = lh$ (b) $A = \frac{1}{2}bh$ (c) $A = 2\pi r h$ (d) $A = \pi r l + \pi r^2$

2   The time of oscillation of a simple pendulum is given by $T = 2\pi \sqrt{(l/g)}$. The mean values of l and g are 3 cm and 981 cm/s^2 and their standard

deviations 0·01 and 0·1, respectively. Calculate the mean value and standard deviation of T.

3   The electrical resistance of a wire is given by $R = E/C$. If the mean values of E and C are 220 and 2 and their standard deviations 1 and 0·1, respectively, calculate the mean and standard deviation of R.

CHAPTER 15

Sampling II

Estimators

In the last chapter we were given the population mean and variance and from them obtained information about sample means and variances. Often the population parameters (population mean, variance, etc.) are not known and estimates of them must be made from samples. If the mean of the sampling distribution of a statistic equals the corresponding population parameter, the statistic is called an unbiased estimator of the parameter; otherwise it is called a biased estimator. The corresponding estimates are called unbiased or biased estimates.

Efficient Estimate of the Population Mean

We showed that the mean of the distribution of sample means was equal to the mean of the population. Therefore the sample mean is an unbiased estimate of the population mean. If two sampling distributions have the same mean the one with the smaller variance is called the efficient estimator of the population mean. We showed that the mean of the means of samples of size n was equal to the population mean, and the variance was equal to σ^2/n where σ is the population standard deviation. It can also be shown that the mean of the medians of samples of a normal, or nearly normal, population is also equal to the population mean, and if n is large its variance is $\sigma^2\pi/2n$ and therefore its standard deviation is $1\cdot2533\sigma/\sqrt{n}$. Therefore the variance of the sample distribution of means is smaller than the variance of the sample distribution of the medians; hence the sample mean is a more efficient estimate of the population mean.

Efficient Estimate of the Population Variance

It can be shown that the mean of the sampling distribution of variances for a large population (or sampling with replacement) is $\sigma^2(n-1)/n$ (see Mathematical Note 15.1) where σ^2 is the population variance and n the number of items in the sample. Thus the sample variance s^2 is not equal to the population variance and is therefore a biased estimate of the population variance σ^2. If we multiply the sample variances by $n/(n-1)$ we then have the mean of the statistics $s^2 n/(n-1)$ equal to σ^2. Therefore $s^2 n/(n-1)$ is an efficient estimator of the population variance. Thus if the estimate of the population variance is denoted by σ_e^2 we have

$$\sigma_e^2 = \frac{ns^2}{n-1} \tag{15.1}$$

Since
$$s^2 = \frac{1}{n} \sum_1^n (x-m)^2$$

where m is the sample mean

$$\sigma_e^2 = \frac{1}{n-1} \sum_1^n (x-m)^2$$

Thus we get an unbiased estimate of σ^2 if we divide the sum of the squared deviations from the sample mean by $(n-1)$ instead of n. In the case of large values of n $(n \geqslant 30)$ $n/(n-1) \simeq 1$ and therefore

$$\sigma_e^2 = ns^2/(n-1) \simeq s^2$$

So in the case of large samples we only incur a small error if the sample variance is taken as an estimate of the population variance.

Confidence Interval Estimates of Population Parameters

If μ and σ are the mean and standard deviation of a normal distribution, then 95% of the area under the curve lies between the ordinates $\mu \pm 1 \cdot 96\sigma$. Therefore we say with 95% *confidence* that a random item of the distribution lies between $\mu \pm 1 \cdot 96\sigma$. Similarly we say with 99% confidence that a random item lies between $\mu \pm 2 \cdot 58\sigma$. We could take other values. Thus the 90% confidence limits are $\mu \pm 1 \cdot 645\sigma$. The 50% confidence limits are $\mu \pm 0 \cdot 6745\sigma$.

In a converse manner if given a random item, x, of a normal distribution we can expect to find the mean of the distribution in the interval $x \pm 1 \cdot 96\sigma$ about 95% of the time, and so on. These are the confidence limits, or fiducial limits, for estimating the mean of the distribution.

EXAMPLE

The mean and standard deviation of the height of a random sample of

100 students is 168·75 and 7·5 cm, respectively. Calculate the (i) 90%, (b) 95%, (c) 99% confidence intervals for estimating the mean height of the whole population of students.

SOLUTION

The best estimate of the population mean from the sample mean is the mean of the sample, that is 168·75. The best estimate of the population variance of the means is the sample variance divided by $(n-1)$, that is

$$\sigma_m^2 = \frac{(7\cdot5)^2}{100-1} \simeq \frac{(7\cdot5)^2}{100}$$

The standard deviation of the distribution of sample means is therefore 0·750 cm.

(i) The 90% confidence limits for the mean of the distribution of sample means which equals the mean of the parent population is

$$168\cdot75 \pm 1\cdot645 \times 0\cdot75 = 168\cdot75 \pm 1\cdot23 \text{ cm}$$

(ii) The 95% confidence limits are

$$168\cdot75 \pm 1\cdot96 \times 0\cdot75 = 168\cdot75 \pm 1\cdot47 \text{ cm}$$

(iii) The 99% confidence limits are

$$168\cdot75 \pm 2\cdot58 \times 0\cdot75 = 168\cdot75 \pm 1\cdot94 \text{ cm}$$

Finite Populations

If the population of students was finite and sampling is without replacement, instead of using $\sigma/\sqrt{n}$ as a measurement of the standard deviation of the distribution of sampling means we should have to use

$$\frac{s}{\sqrt{n}}\sqrt{\left(\frac{N-n}{N-1}\right)}$$

Thus, if the size of the parent student population is 2000

$$\sigma_m = \frac{7\cdot5}{\sqrt{100}}\sqrt{\left(\frac{2000-100}{2000-1}\right)} = 0\cdot731$$

which makes the 90% confidence limits

$$168\cdot75 \pm 1\cdot645 \times 0\cdot731 = 168\cdot75 \pm 1\cdot20$$

EXAMPLE

In a large school the number of students sitting O level G.C.E. was 240. A random sample of 60 students was selected and given a mock examination. Their mean percentage mark was 48 with a standard deviation of 16. Assuming the standard in the O level examination of the question paper and marking will be equal to those in the mock examination what are the 95% confidence limits for estimates of the mean mark of the 240 students in the O level examination?

SOLUTION

The population size is not infinite. The estimate of the standard deviation of the means of samples of size 60 is

$$\frac{16}{\sqrt{60}}\sqrt{\left(\frac{240-60}{240-1}\right)} = 0\cdot567$$

The 95% confidence limits are

$$48\pm1\cdot96\times0\cdot567 = 48\pm1\cdot11$$

EXAMPLE

The standard deviation of the diameters of ball-bearings is 0·04 cm. How many ball-bearing diameters must be measured so that we have 95% confidence limits that our estimate of the mean of the population does not differ from the true mean by more than 0·01 cm?

SOLUTION

The 95% confidence limits are $\mu_m\pm1\cdot96\sigma/\sqrt{n}$. Therefore $1\cdot96\sigma/\sqrt{n}$ or $0\cdot04\times1\cdot96/\sqrt{n} = 0\cdot01$. Hence $n = 61\cdot47$ so the size of the sample must be 62.

EXAMPLE

A sample of 200 packets of a product supplied by one manufacturer has a mean mass of 8·2 kg with a standard deviation of 0·45 kg. A sample of 300 packets of the same product from another manufacturer has a mean mass of 7·8 kg with a standard deviation of 0·3 kg. Calculate the

80% confidence limits for the difference in the mean weights of the populations of the products.

SOLUTION

The variance of the means of samples from the first manufacturer is $(0.45)^2/200$ and from the second manufacturer $(0.3)^2/300$. Therefore the variance of the difference of the means is

$$\frac{(0.45)^2}{200} + \frac{(0.3)^2}{300} = 0.004$$

Therefore the standard deviation of the difference of the means is 0.02. 80% of the area below the normal curve is included between the ordinates at ± 1.28. Therefore the 80% confidence limits for the difference in the mean masses of the products are

$$(8.2 - 7.8) \pm 1.28 \times 0.02 = 0.4 \pm 0.0256$$

That is four times out of five we can be confident the difference in the mean masses of samples of the products will lie between 0.374 and 0.426 kg.

Exercise 15.1

1 The length of an object was measured 50 times and the distribution of the measurements was as follows

Length (cm)	21.53	21.54	21.55	21.56	21.57	21.58
Frequency	3	10	16	12	6	3

Calculate the 99% confidence limits for the length of the object.

2 The mean and standard deviation of the breaking strengths of 100 rods is 300 N and 2.3 N, respectively. Estimate the (i) 99%, (ii) 95%, (iii) 60% confidence limits for the breaking strengths of all rods produced by the company.

3 The standard deviation of the lifetime of an article is 144 hours. How large a sample must be taken to be (a) 95%, (b) 99% confident that the error in the estimated mean lifetime of the article will not exceed (i) 15 hours, (ii) 20 hours?

4 Two lots of electric light bulbs were tested for length of life. Using the data given below can we place any confidence in their being equally good?

	Size of sample	Mean life (h)	Standard deviation (h)
Lot A	200	2000	400
Lot B	250	2120	450

5   The hourly output of a certain machine has a mean of 650 units and a standard deviation of 45 units. With a new operator a sample of ten one-hour test runs gave a mean of 700 units. Does the result justify the assumption that the new operator is more efficient?

What are the 95% confidence limits for the mean hourly rates for the new operator?

6   The fat content of a product was measured by a manufacturer and a purchaser. The manufacturer made 80 tests and found the mean fat content as 13·108 with a standard deviation of 0·014. The purchaser made 65 tests and obtained a mean of 13·101 with a standard deviation of 0·016. Comment on the results.

7   The table gives the lives of electric light bulbs in a sample of 983. Determine the 90% confidence limits for the mean life of the lamps.

Life (mid-value in hours)	200	400	600	800	1000	1200	1400	1600	1800
Frequency	1	3	35	54	126	170	205	180	113
	2000	2200	2400	2600	2800	3000	3200	3400	
	60	27	6	1	0	1	0	1	

8   Daily tests of a product were taken by a firm for 1406 days and the mean result was 545·86 with a variance of 15·233. During this time an inspector took 72 spot checks and the mean value of his checks was 546·92 with a variance of 27·922. Is there any evidence of a difference between the two results?

Other Common Statistics

We have used the mean and standard deviation to illustrate the methods under discussion. These methods can be applied equally well to other statistics.

(i) Standard Deviations
If a population is normal, or nearly normal, the standard deviation of sample standard deviations is $\sigma/\sqrt{(2n)}$.

If $n \geqslant 100$ the distribution of the sample standard deviations is approximately normal, and their mean is nearly equal to the population standard deviation.

(ii) Proportions
If p is the probability of success in a binomial distribution it was pointed

out in *Elementary Statistics* page 256 that the proportion of successes expected in a sample of size n is p with a standard deviation $\sqrt{(pq/n)}$.

For large samples we can use the sample estimate, p_s, for p and the 90% confidence limits for the population proportion of successes are

$$p_s \pm 1 \cdot 96 \sqrt{\left(\frac{p_s q_s}{n}\right)}$$

if sampling is from an infinite population, and

$$p_s \pm 1 \cdot 96 \sqrt{\left(\frac{p_s q_s}{n}\right)} \sqrt{\left(\frac{N-n}{N-1}\right)}$$

if sampling is without replacement from a population of size N.

(iii) Medians
The mean of sample medians of normal populations is equal to the mean of the population, and for large samples the standard deviation of the sample medians is

$$\sigma \sqrt{\frac{\pi}{2n}} = \frac{1 \cdot 2533 \sigma}{\sqrt{n}}$$

(iv) Quartiles
The means of the first and third quartiles of samples from normal populations are nearly equal to the first and third quartiles of the population, and the standard deviations of the sample quartiles are equal, and for large samples are nearly normally distributed. The standard deviation is $1 \cdot 3626 \sigma / \sqrt{n}$.

(v) Semi-interquartile Range
The remarks made for quartiles apply to the semi-interquartile range. The mean of sample inter-quartile ranges is nearly equal to the population semi-interquartile range and the standard deviation is $0 \cdot 7867 \sigma / \sqrt{n}$.

(vi) Variances
The remarks made for standard deviations apply to variances. For large samples the mean of sample variances is equal to the population variance and their standard deviation is

$$\sigma^2 \sqrt{\frac{2}{n}}$$

EXAMPLE

A random sample of 700 apples from a large consignment showed that

210 were bruised. Find 95% confidence limits for the proportion of bruised apples in the consignment.

SOLUTION

The probability of finding a bruised apple is $p = 210/700 = 0.3$. Therefore $q = 1-0.3 = 0.7$. Therefore the 95% confidence limits for the proportion of bruised applies is

$$0.3 \pm 1.96 \sqrt{\left(\frac{0.3 \times 0.7}{700}\right)} = 0.3 \pm 0.034$$

EXAMPLE

The standard deviation of a large population is 12·0 units. Samples of size 100 are drawn and the standard deviation of each sample calculated. Calculate (i) the mean, (ii) the standard deviation of the sampling distribution of sample standard deviations.

SOLUTION

(i) The mean of the sampling distribution of standard deviations equals the standard deviation of the population, that is 12·0 units.
(ii) The standard deviation of the sampling distribution of standard deviations equals

$$\frac{\sigma}{\sqrt{2n}} = \frac{12.0}{\sqrt{200}} = 0.853$$

EXAMPLE

The median of a population is 800 and its standard deviation is 60. If we take 2000 samples of size 100, in how many samples would we expect the median to exceed 810?

SOLUTION

The mean of the sample medians should equal 800 and the standard deviation of the sample medians is

$$\frac{1.2533\sigma}{\sqrt{n}} = \frac{1.2533 \times 60}{10} = 7.5198$$

Therefore the standardized normal variate is

$$\frac{810-800}{7 \cdot 5198} = 1 \cdot 33$$

and the probability of exceeding this variate is $0 \cdot 0918$. Therefore the number of sample medians expected to exceed 810 is $0 \cdot 0918 \times 2000 = 184$.

Decision Theory

We have shown how the methods of statistics can be used for estimating the values of specific statistics; we will now show how they can be used to decide the 'truth' or otherwise of a given hypothesis. The question 'Does smoking increase the chances of developing lung cancer?' was decided by statistical methods.

Let us consider a simple case. We are given a coin and asked to decide if it is biased. The first thing to do is to make a hypothesis, called the null hypothesis, and the second is to perform an experiment to test its truth or otherwise; in other words, to try to nullify it. In our case the simplest null hypothesis is to assume the coin is not biased. The simplest experiment to test the truth of the hypothesis is to toss the coin a number of times and count the number of heads or tails and see if the number agrees with what we expect if the hypothesis is true.

If the coin is unbiased the probability of tossing a head is $\frac{1}{2}$ and in 100 tosses the mean number of heads expected is $100 \times \frac{1}{2}$, or 50, and the standard deviation is $\sqrt{(100 \times \frac{1}{2} \times \frac{1}{2})}$ or 5. The distribution of the number of heads will be normal, and the error introduced by considering it as continuous is negligible. If the variates are standardized the distribution can be represented by the standardized normal curve. The standardized variate has a 95% chance of falling between A and B (Diagram 15.1) and a 5% chance of falling in the shaded area below A or above B. Let us decide that if the variate falls between A and B the evidence suggests the event can quite easily happen in random sampling and we have no reason to doubt the hypothesis, but if the standard variate falls below A or above B then we consider the event to be rare in random sampling and it throws doubt on the truth of the hypothesis.

Suppose the experiment produced 56 heads. Standardizing this variate we have (remembering that the number of heads is a discrete variable and is measured on the continuous curve by the interval 55·5 to 56·5), $(56 \cdot 5 - 50)/5 = 1 \cdot 3$, and this falls between A and B. Therefore we have no evidence to suggest the coin is biased. If the experiment produced 38 heads the standardized variate is $(37.5-50)/5 = -2.5$, and this variate falls below A. We have to take the lower value 37·5 because it is the outer value of the interval whose mid-point is 38 just as previously 56·5 was the outer limit of the interval whose mid-point is 56. We have decided that this event is too rare to occur by chance and conclude that the evidence suggests the coin is biased. It is necessary to make this a 'two-tailed' test because the appearance of too many tails indicates bias equally as well as the appearance of too many heads.

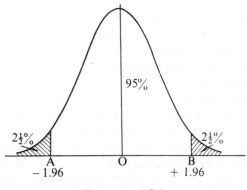

DIAGRAM 15.1

Type I Errors

Anyone is justified in saying that an event with a probability of 1/20 (i.e. 5%) is likely to occur at times, and that horses whose odds are 100 to 1 sometimes win. Therefore, rejecting an event whose normal standardized variate falls in the shaded area may be a mistake. This is known as a Type I error, and is the error we make by rejecting an event as unlikely when we should have accepted it. We could reduce the possibility of this type of error by reducing the size of the shaded area in Diagram 15.1.

Type II Errors

Suppose the coin was biased and the probability of obtaining a head was 0·7 and we did not know. If we tossed the coin 100 times and took a large number of such samples, we would obtain a normal distribution whose mean was $100 \times 0.7 = 70$ and standard deviation $\sqrt{(100 \times 0.7 \times 0.3)} = 4.583$. If this and the previous distribution are drawn in the same diagram, we get the two normal curves as in Diagram 15.2. These two curves overlap and we see that the appearance of 40 to 60 heads could happen with the biased coin. This introduces another type of error—accepting a hypothesis (i.e. $p = 0.5$) when it should be rejected. To reduce the probability of this error we should move the points A and B nearer to the mean 50, and reduce the area of curve 2 between these points. This introduces a dilemma because reducing the probability of making a Type II error increases the probability of making a Type I error, and conversely. So, when fixing our significance values we must consider which error produces the more serious consequences and reduce the probability of making it.

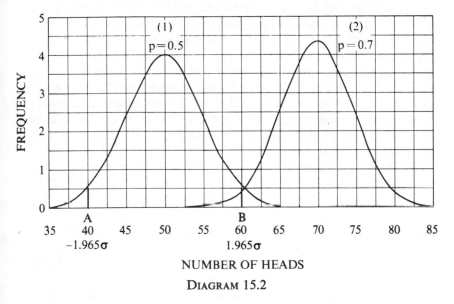

NUMBER OF HEADS

DIAGRAM 15.2

The probability of making a Type II error in the example given is the area under the normal standardized curve for curve 2 between the variates 40 and 60. Standardizing these variates we have $(39.5 - 70)/4.583$, that is -6.66, and $(60.5 - 70)/4.583$, that is -2.07. From the tables this gives us a probability of 0.0192. So the probability of making a Type II error, that is of accepting $p = 0.5$ when $p = 0.7$, occurs about once in every 50 experiments. (Note: the error that would have been introduced by neglecting the area to the left of A would have been negligible.)

Choice of Error

The choice of probability for significance purposes depends on which error is likely to produce the more serious results. If we wish to avoid Type I errors we could reduce the probability significance. In our example we took a probability of 0·05. If we took 0·01 we would reduce Type I errors, but increase Type II errors. Suppose Type II errors are the more important to avoid, we could employ a significance probability of 0·10 or more, but this would increase the probability of a Type I error.

Let us illustrate the point. A school uses a number of expensive pieces of laboratory equipment with a short life, and a manufacturer claims to be able to manufacture a new piece of equipment with a longer life than the old equipment but it is more expensive than the old equipment. The master in charge of the laboratory decides to compare the life of the new equipment with that of the old. If he makes a Type I error, that is, rejecting the old equipment and accepting the new, the new equipment will be installed and a large amount of money wasted when the new equipment is not better than the old. If a Type II error is made, that is, he rejects the new equipment when it has a longer life, the new equipment is not installed. This may be a disadvantage, but no positive loss of money has occurred. The master will probably decide it is more important to avoid a Type I error than a Type II.

Let us take a second illustration. A potent drug has been in use for some time and its properties have been well tested. A manufacturer claims to have discovered a new drug which is better than the old. If, when the new drug is compared with the old drug, a Type I error is made, the drug will be rejected although it is really better than the old. On the other hand if a Type II error is made and the drug is accepted and used on patients it may produce serious side effects, perhaps resulting in death. In this case it is more important to avoid Type II errors.

It is difficult to make generalizations about which error to avoid and the investigator must decide for himself according to the problem under investigation. Time should be spent before the test is made on the probability to be used. This avoids wishful thinking on the part of the investigator.

Power Curves

If the experiment with the coin is repeated for different values of p, the probability of making a Type II error, that is, of accepting the hypothesis $p = 0.5$ with a 5% level of significance when p has a different value, is tabulated below:

p	0·1	0·2	0·3	0·4	0·5	0·6	0·7	0·8	0·9
β	0·0000	0·0000	0·0192	0·5040	0·9642	0·5040	0·0192	0·0000	0·0000

β is called the *operating characteristic function* of a test and is the probability of making a Type II error. In our test it is the probability of making an error in samples of 100 at a 5% significance level, that is, of accepting the hypothesis $p = 0.5$ when the correct value of p is given in the first row. Thus, the probability of accepting the hypothesis $p = 0.5$ when the correct value of p is 0·7, is 0·0192. The probability of accepting $p = 0.5$ when p is 0·5 is 0·9642, and so on.

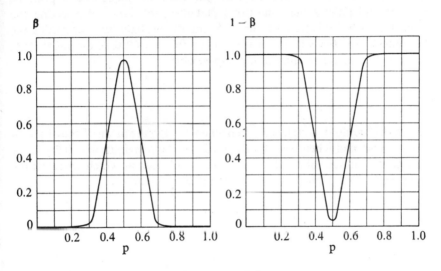

DIAGRAM 15.3

$(1 - \beta)$ gives us a measure of the probability of rejecting the hypothesis, say of rejecting $p = 0.5$, when the correct hypothesis is $p = 0.7$. Thus, the probability of rejecting the hypothesis $p = 0.5$, when $p = 0.2$, is $1 - 0.0000 \simeq 1$ (0·0000 is given to 4 decimal places). The probability of $p = 0.5$ being rejected is $1 - 0.5040 = 0.4960$, when $p = 0.4$.

The value of $(1 - \beta)$ is called the *power function* since it indicates the *power of the test* to reject a wrong hypothesis. The graph of β against p is called the *operating characteristic* or OC curve, and the graph of $(1 - \beta)$ against p (which is the inverse of the first curve) is called the *power curve*.

Sample Sizes

The last test gave us the probability of accepting or rejecting a test for a sample size of 100. Suppose we now lay down the acceptable probabilities of making Type I or Type II errors and calculate the minimum sample size necessary.

EXAMPLE

If in the previous experiment with the coin we make the hypothesis $p = 0.5$ and impose the conditions that the probability of rejecting the hypothesis when it is true is to be 0.05, and the probability of accepting the hypothesis $p = 0.5$ when p really equals 0.7 must be at the most 0.06, what is the necessary minimum sample size?

SOLUTION

If n is the required sample size then when $p = 0.5$ the mean of the distribution is $0.5n$, and when $p = 0.7$ the mean of the distribution is $0.7n$. Let x denote the number of heads in n tosses of the coin at which we reject the hypothesis. The problem can be illustrated as follows.

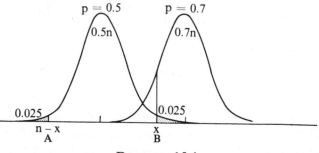

DIAGRAM 15.4

Since the distribution $p = 0.5$ is symmetrical about its mean $0.5n$ and the value of the number of heads at B is x, the value of the variable at A is $(n - x)$, for the arithmetic mean of $(n - x)$ and x is $0.5n$. We can now proceed with the solution of the problem. To make a Type I error the value of the dark shaded area of the curve $p = 0.5$ to the right of B in the standardized normal curve must equal 0.025. Therefore,

$$\frac{x - np}{\sqrt{(npq)}} = \frac{x - 0.5n}{\sqrt{[n(0.5)(0.5)]}} = 1.96 \quad \text{(from Table 2)}$$

To make a Type II error the area of the standardized normal curve for $p = 0.7$ between A and B must be 0.6.

The standardized variable at $B = \dfrac{x - 0.7n}{\sqrt{[n(0.7)(0.3)]}}$

and at A it is $\dfrac{(n - x) - 0.7n}{\sqrt{[n(0.7)(0.3)]}}$

and the difference between these areas must be 0.6. There is no way of calculating these areas so we make the assumption that the total area of the standardized curve for $p = 0.7$ to the left of B is 0.06 and neglect the area to the left of A. Now

$$\frac{x - 0.7n}{\sqrt{[n(0.7)(0.3)]}} = -1.555 \quad \text{(from Table 2)}$$

Reducing both the equations, we have

$$x - 0.5n = 0.9800\sqrt{n}$$
$$x - 0.7n = -0.7127\sqrt{n}$$

Subtracting and solving for n we have

$$n = 71.64 \simeq 72 \text{ tosses per sample}$$

The value of $x = 44.116 \simeq 45$. If $n = 72$ and $p = 0.5$ the mean of the distribution is 36. So $x - np = 45 - 36 = 9$.

So we accept the hypothesis under the conditions laid down if the number of heads in samples of size 72 is in the range 36 ± 9.

We can now show that the error introduced by neglecting the area to the left of A in the standardized curve for $p = 0.7$ is negligible.

The standardized variate at A is

$$\frac{(n - x) - np}{\sqrt{(npq)}} = \frac{27.524 - 71.64 \times 0.7}{\sqrt{(71.64 \times 0.7 \times 0.3)}} = -5.832$$

The magnitude of this standardized variate is too large for our tables, and we can neglect the area to the left of A in the degree of accuracy required for our problem.

(Using special tables the area is $1 - 0.9999999972 = 0.0000000028$)

Exercise 15.2

1 A firm has been manufacturing fibres with a mean breaking strength

of 200 N and a standard deviation of 20 N. A research worker claims he has developed a process which will increase the mean breaking strength.

(a) What should be the decision rule for rejecting the old process at a 0·01 level of significance if it is decided to test 100 fibres?

(b) Under the decision rule in (a) what is the probability of accepting the old process when the new process has increased the mean breaking strength to 210 N without a change in the standard deviation?

2   Construct an OC curve and a power curve for the test in question 1 for values of the new mean of 195, 200, 205, . . ., 225. (In this case it is easier to plot μ against p.)

3   The probability of recovering from a certain illness with the aid of a given drug is 0·7. A manufacturer claims to have discovered a new drug where the probability of recovery is 0·8. In an experiment to test the validity of the manufacturer's claim it is assumed that the old drug is the better and the probability of rejecting this drug in random samples must be 0·05 at most, and the probability of accepting the old drug when the manufacturer's claim may be true must be 0·20 at most.

Calculate the necessary minimum sample size, and state the decision rule to be made.

4   A bag is assumed to contain equal numbers of red and black coloured balls. If sampling is with replacement, what is the probability of rejecting the hypothesis if it is to be accepted only if between 44 and 56 red balls are drawn in samples of 100?

Determine the probability of accepting the hypothesis in the first part of the question when the proportion of red balls is 6/10.

Repeat the above calculation if the proportion of red balls is (i) 7/10, (ii) 8/10, (iii) 9/10.

Draw an OC curve and a power curve.

Mathematical Note 15.1.   The Mean of Variances of Samples.

Let s^2 denote the mean square deviation of the values of the items of a sample from their own mean m, i.e. s^2 is the variance of the sample, and

$$s^2 = \frac{1}{n}\sum_{1}^{n} x^2 - m^2$$

As in Mathematical Note 14.1 the number of times each x appears is

$_{N-1}C_{n-1}$. So summing over all samples, we have, if N is the total number of x's in the population

$$\Sigma s^2 = \frac{1}{n} \, _{N-1}C_{n-1} \sum_1^N x^2 - \Sigma m^2$$

There are $_NC_n$ samples; so the mean value of s^2 is

$$\frac{\Sigma s^2}{_NC_n} = \frac{1}{N} \sum_1^N x^2 - \frac{\Sigma m^2}{_NC_n}$$

In Mathematical Note 14.1 we proved (see Equation (5)) that

$$\sigma_m^2 = \frac{\Sigma m^2}{_NC_n}$$

Therefore the mean value of s^2 is

$$\frac{1}{N} \sum_1^N x^2 - \sigma_m^2$$

$$= \sigma^2 - \frac{\sigma^2}{n}\left(\frac{N-n}{N-1}\right) \quad \text{(Mathematical Note 14.1 Equation (6),}$$
$$\text{and mean of population} = 0)$$

$$= \sigma^2 \frac{N(n-1)}{n(N-1)} \qquad \text{for all populations}$$

$$= \sigma^2 \frac{(n-1)}{n\left(1-\dfrac{1}{N}\right)} \qquad \text{(dividing numerator and denominator by } N)$$

$$= \sigma^2 \frac{(n-1)}{n} \qquad \text{if } N \text{ is large}$$

$$= \sigma^2 \qquad\qquad \text{if } n \text{ is also large}$$

CHAPTER 16

Small Samples and the t-Distribution

Exact Sampling

So far in sampling theory we have made use of the fact that for samples of size greater than 30 items, called large samples, the sampling distributions of many statistics are approximately normal, and the approximation improves with an increase in the size of the sample. For samples of less than 30 items, called small samples, the sampling distributions of the statistics are not approximately normal, and become less so for very small samples. In this case we must be more exact and the approximations we make in the theory of sampling for n being large can no longer be made. Small sampling theory is exact sampling theory and holds for large as well as small samples.

t-Distribution

For observations taken from a normal, or nearly normal distribution the distribution of the statistics of small samples has the form

$$y = y_0\left(1+\frac{t^2}{v}\right)^{-\frac{1}{2}(v+1)}$$

where t takes the place of the standardized normal variate and is equal to $(x-\mu)/\sigma_e$, where x is a variate, μ the population mean and σ_e^2 an unbiased estimate of σ^2, the population variance. v is the number of degrees of freedom available for estimating σ^2, and y_0 is a constant chosen so that the area under the curve is unity. The formula represents a family of curves, one for each value of v, and like the normal curve each curve is symmetrical about the y-axis and asymptotes to each end of the t-axis.

Diagram 16.1 shows the characteristics of this family of curves, and how decreasing values of the magnitude of v increases the spread. Areas under these curves have been calculated and a separate table should be used for

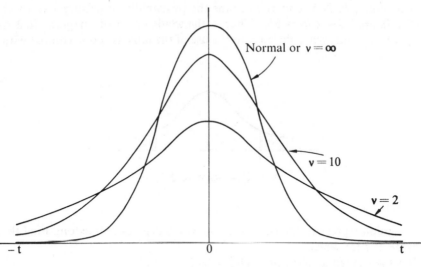

DIAGRAM 16.1

each value of v. The tables give the probability of obtaining a given value of t and are used in a similar way to the normal table.

Forms of t Table

To save using bulky tables various forms of abbreviated t-tables are published and it is a time-saver if the form of the table is first studied. In Table 3 p. 336, percentage points of the t-distribution are given and the table is read as follows. For $v = 4$ and $t = 2.78$ we see the value of the

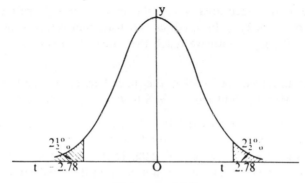

DIAGRAM 16.2

probability, P, is 5, and this means the probability of getting a value of $t \geqslant 2 \cdot 78$ and $\leqslant -2 \cdot 78$ is 5%. That is, the shaded area of Diagram 16.2 is 5% of the total area. Efficiency in the use of the table is soon acquired with practice.

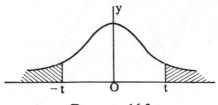

DIAGRAM 16.3

EXAMPLE

The graph represents the t-distribution with 8 degrees of freedom. Find the value of t for which
  (i) the shaded area on the right $= 0 \cdot 05$, i.e. 5%
 (ii) the total shaded area $= 0 \cdot 05$, i.e. 5%
(iii) the total unshaded area $= 0 \cdot 99$, i.e. 99%
(iv) the shaded area on the left $= 0 \cdot 01$, i.e. 1%

SOLUTION

(i) If the shaded area on the right is $0 \cdot 05$ the total shaded area is $0 \cdot 1$ or 10%.
   In the table of the t-distribution proceed down the v column until 8 is reached and then proceed right until under $P = 10$. The result is $1 \cdot 86$, the required value of t.

(ii) If the total shaded area is 5%, the area to the right of t is $2\frac{1}{2}$% and to the left of $-t$ is $2\frac{1}{2}$%. Proceeding as before, that is down the v column to 8 and to the right until we reach the value of t under $P = 5$ we find $t = 2 \cdot 31$.

(iii) If the unshaded area is 99%, the shaded area is 1% and each shaded area is $\frac{1}{2}$%. Moving right from $v = 8$ to $P = 1$% we have $t = 3 \cdot 36$.

(iv) If the shaded area on the left is 1%, then by symmetry the shaded area on the right is 1%, and the total shaded area is 2%. Moving right from $v = 8$ to $P = 2$% we get $t = 2 \cdot 90$. Therefore $-t = 2 \cdot 90$.
   Values of t for values of v and P not given in the table can be found by interpolation.

Exercise 16.1

1 Find the critical values of t if the area of the right-hand tail in Diagram 16.3 is $2\frac{1}{2}\%$ and the number of degrees of freedom equals (i) 15, (ii) 28, (iii) 35, (iv) 250.

2 What are the 95% confidence coefficients (two-tailed) for the t-distribution if v equals (i) 10, (ii) 21, (iii) 36, (iv) 90?

Significance of a Single Mean

Most of the sampling problems deal with the significance of the mean, and this involves estimating the mean and variance of the parent population. One type of problem deals with the significance of a single mean.

In Chapter 15 we showed that the best estimate of the mean of the population is the sampling mean, and the best estimate of the population variance is $ns^2/(n-1)$, where s^2 is the variance of the sample. If we are given the individual values of the items composing the sample, then we have shown

$$\frac{n}{n-1}s^2 = \frac{n}{n-1} \times \frac{\Sigma(x-m)^2}{n} = \frac{\Sigma(x-m)^2}{n-1}$$

The number of degrees of freedom is $n-1$.

EXAMPLE

A sample of 10 measurements of the diameters of ball-bearings gave a mean of 0·445 cm and a standard deviation of 0·006 cm. What are the 95% confidence limits for the mean of the diameters?

SOLUTION

The best estimate of the mean diameter is 0·445 cm, and the best estimate of the variance of ball-bearings is

$$\frac{n}{n-1}s^2 = \frac{10}{10-1}(0·006)^2$$

Therefore the variance of sample means of samples of size 10 is

$$\frac{10}{9}(0·006)^2 \times \frac{1}{10} \qquad \left[\text{i.e. } \frac{\sigma^2}{n}\right]$$

and the estimate of standard deviation of sample means is therefore

$$\frac{0.006}{3} = 0.002 \text{ cm}$$

Therefore the 95% confidence limits for 9 degrees of freedom is

$$0.445 \pm 2.26 \times 0.002$$
$$= 0.445 \pm 0.00452 \text{ cm}$$

EXAMPLE

A machine is supposed to produce washers of thickness 0·14 cm. To find if the machine is in proper working order a sample of 10 washers is taken. The thicknesses are 0·1275, 0·13, 0·1325, 0·1325, 0·135, 0·135, 0·1375, 0·1375, 0·14, 0·1425, cm. Using a level of significance of 5% is the machine in proper working order?

SOLUTION

If x denotes the thickness of a washer, the sample mean

$$= \frac{\Sigma x}{n} = \frac{1.35}{10} = 0.135 \text{ cm}$$

Therefore,

$(x-m)$ $-7.5, -5, -2.5, -2.5, 0, 0,$ $2.5,$ $2.5,$ $5,$ $7.5,$ thousandths cm
$(x-m)^2$ $56.25, 25, 6.25, 6.25, 0, 0, 6.25, 6.25, 25, 56.25$ $\Sigma(x-m)^2 = 187.5,$

and the estimated *population* variance is 187·5/9. The estimated population standard deviation is $\sqrt{(187.5/9)} = 4.565$ thousandths cm $= 0.004565$ cm. Therefore the standard deviation of sample means is

$$\frac{0.004565}{\sqrt{10}} = 0.00145$$

Therefore
$$t = \frac{0.135 - 0.14}{0.00145} = -3.448$$

Since the washers can be too big or too small, this is a two-tailed test and for 9 degrees of freedom and a 5% significance level $t = 2.26$. Therefore the result is highly significant and the machine is not in proper working order.

Paired Samples

In this type of problem subtract the values of each pair of items and the differences should form a distribution with zero mean. Use the *t*-test to see if the sample satisfies such a distribution.

EXAMPLE

Five pupils were given a statistics test. They were given a month's further teaching and a second test of equal difficulty was then held. Their marks in the two tests were

Pupil	A	B	C	D	E
Test 1	20	18	22	19	20
Test 2	22	19	19	18	23

Do these marks give evidence of improvement from the extra tuition?

SOLUTION

The problem really asks if the mean of the differences between each pair of marks differs significantly from zero. If x is the difference between pairs of marks, then

$$x = -2, -1, 3, 1, -3$$

The mean $= -2/5 = -0.4$.

$(x-m) = -1.6, -0.6, 3.4, 1.4, -2.6$

$(x-m)^2 = 2.56, 0.36, 11.56, 1.96, 6.76$, and $\Sigma(x-m)^2 = 23.20$

The estimated population standard deviation $= \sqrt{(23.20/4)} = 2.409$.

The estimated standard deviation of sample means $= 2.409/\sqrt{5} = 1.077$.

Therefore

$$t = \frac{-0.4-0}{1.077} = -0.3764$$

From the *t*-tables for $v = 4$ we see that this value of t is quite likely to occur and we conclude there is lack of evidence to show any difference in the performance of the pupils.

Unpaired Samples

In this case we have two samples which are not necessarily equal in size and we wish to know if the difference of their means is significant.

Let the two samples be of size n_1 and n_2, and let their means be m_1 and m_2. The null hypothesis is that both samples are random samples from the same normal population of unknown mean, μ, and variance σ^2.

The distribution of the mean of the first sample, m_1, is normal of mean μ and standard deviation $\sigma/\sqrt{n_1}$, and of the second sample, m_2, also of mean μ and standard deviation $\sigma/\sqrt{n_2}$. Therefore $m_1 - m_2$ is normal variate of a distribution whose mean is zero and standard deviation

$$\sqrt{\left(\frac{\sigma^2}{n_1} + \frac{\sigma^2}{n_2}\right)} = \sigma\sqrt{\left(\frac{n_1 + n_2}{n_1 n_2}\right)}$$

It may seem that the best way to obtain an estimate of σ^2, the population variance, is to combine the two samples and estimate σ^2 from this larger sample, but for mathematical reasons it is wrong to combine the two samples and to use the combined sample variance to estimate σ. It is better to proceed as follows.

If $s_1{}^2$ and $s_2{}^2$ are the two sample variances then

$$n_1 s_1{}^2 \text{ is an estimate of } (n_1 - 1)\sigma^2$$
and $\quad\quad n_2 s_2{}^2$ is an estimate of $(n_2 - 1)\sigma^2$

Adding $n_1 s_1{}^2 + n_2 s_2{}^2$ is an estimate of $(n_1 + n_2 - 2)\sigma^2$.

Therefore an estimate of σ^2 is

$$\sigma_e{}^2 = \frac{n_1 s_1{}^2 + n_2 s_2{}^2}{n_1 + n_2 - 2} \tag{16.1}$$

with $n_1 + n_2 - 2$ degrees of freedom.

This can be extended to more than two samples. Thus the best estimate of the population variance from k samples is given by

$$\frac{n_1 s_1{}^2 + n_2 s_2{}^2 + \ldots n_k s_k{}^2}{n_1 + n_2 + \ldots n_k - k}$$

EXAMPLE

The I.Q.'s (intelligence quotients) of 12 students from one school showed a mean of 108 with a standard deviation of 11, while the I.Q.'s of 10 students of the same age-group from another school in the same town had a mean of 113 with a standard deviation of 9. Is there a significant difference between the I.Q.'s of the two groups?

SOLUTION

$m_1 = 108; s_1 = 11; n_1 = 12; m_2 = 113; s_2 = 9; n_2 = 10$

$$\sigma_e^2 = \frac{n_1 s_1^2 + n_2 s_2^2}{n_1 + n_2 - 2} = \frac{12 \times 11^2 + 10 \times 9^2}{12 + 10 - 2} = 113 \cdot 1$$

$$\sigma_e = 10 \cdot 63$$

Therefore the standard deviation of the difference of the means

$$= \sigma_e \sqrt{\left(\frac{n_1 + n_2}{n_1 n_2}\right)} = 10 \cdot 63 \sqrt{\left(\frac{12 + 10}{120}\right)} = 4 \cdot 551$$

Therefore $\qquad t = \dfrac{113 - 108}{4 \cdot 551} = 1 \cdot 099$

Therefore the difference in the means is not significant.

Exercise 16.2

1 Four different boxes of matches packed by the same machine contain 53, 57, 52 and 54 matches. Obtain 95% confidence limits for the mean number of matches packed in boxes by the same machine.

2 The mass in kilogrammes of a certain product was measured for 12 samples. The results were 1·78, 1·79, 1·84, 1·85, 1·83, 1·79, 1·86, 1·82, 1·85, 1·74, 1·77, 1·80. Calculate 90% confidence limits for the mean mass of the product.

3 A machine is supposed to produce paper with a mean thickness of 0·005 cm. Nine random measurements of the paper have a mean of 0·00470 cm with a standard deviation of 0·00030 cm. Is the output significantly different with respect to thickness from standard at the 5% level?

4 A random sample of the heights of nine boys were 120, 110, 135, 125, 147·5, 145, 147·5, 150, 157·5 cm. Is the mean height significantly different from 125 cm at the 5% level?

5 To test the effect of a fertilizer on the growth of a plant ten seedlings were grown in each of two boxes under exactly similar conditions except that the seedlings in one box were treated with the fertilizer. The differences

in heights of the paired plants when fully grown were measured and the following results obtained. Fertilized minus Unfertilized (cm) 6·1, 2·3, 10·2, 23·8, 3·4, 5·2, −1·6, −15·2, −3·7, 10·8.

Does the experiment provide satisfactory evidence of the success of the fertilizer?

6   Three athletes are coached to run the 100 metres. Before coaching their times are 10·8, 10·9, and 11·0 sec and after coaching 10·5, 10·6 and 10·6 sec, respectively. Does this prove the coaching effective?

7   A random sample of 10 items has a mean of 112 and a standard deviation of 6·6. Could this sample have come from a population whose mean is 104?

8   A group of six adults was tested to find out how many words they could repeat from memory after reading from a list. After the lapse of a week they were retested. Is the difference in the performance of the six adults significant?

Adult	A	B	C	D	E	F
Test 1	6	4	7	5	8	6
Test 2	7	6	7	7	9	6

9   Each of nine pieces of nylon thread is divided into two equal pieces and one of each pair is treated with a chemical and the other is left untreated. The breaking strengths of each of the nine pairs are shown below. Is there a significant difference between the mean breaking strengths of the treated and untreated thread?

Pair	1	2	3	4	5	6	7	8	9
Treated	281	331	334	317	315	353	359	296	328
Untreated	305	326	315	317	361	318	345	286	323

10   The test marks of ten students before and after being coached in a subject were:

Student	A	B	C	D	E	F	G	H	I	J
Marks before	43	49	50	50	30	46	65	40	70	50
Marks after	50	47	57	44	52	62	60	36	83	65

Was the coaching effective?

11   The marks of six boys and four girls in a test were:

Boys	74	43	82	40	67	66
Girls	78	88	72	62		

Is the difference significant?

12 Two samples of a plant were measured for height with the following results.

First sample (cm)	182	169	171	183	205	171	168	
Second sample (cm)	163	184	184	177	158	176	166	181
	181	168	175	150				

Find 95% confidence limits for the difference in the mean heights for the two populations.

13 Samples of twelve and fifteen animals were fed on different diets A and B, respectively. The gains in weight for the individual animals for the same period were as follows:

| Diet A newton | 20 | 25 | 23 | 29 | 19 | 30 | 8 | 27 | 19 | 25 | 26 | 30 | | | |
| Diet B newton | 39 | 29 | 17 | 13 | 42 | 26 | 35 | 25 | 27 | 30 | 13 | 16 | 30 | 24 | 17 |

Assuming the diet affects the mean gain without affecting the variance of gains, calculate 90% confidence limits for the average increase in weight of animals on diet B as compared with those on diet A.

14 Five pairs of identical and adjacent plots were used to test the effect of two fertilizers A and B on the production of corn. One plot of each pair was treated with fertilizer A and the corresponding plot of each pair with fertilizer B. The plots with fertilizer B yielded 27, 8, 6, 5 and 4 litres per are more than the corresponding plots treated with fertilizer A. Is fertilizer B demonstrated as superior to fertilizer A? (1 are = 100 m^2)

15 The barometric pressures in millibars at two observatories A and B, taken at random times, are as follows:

| A | 1002 | 996 | 998 | 1027 | 1031 | 1015 | 991 |
| B | 972 | 984 | 987 | 999 | 979 | 965 | |

Examine the hypothesis that the mean barometric pressure is the same at A as at B.

Suggest a better way of making observations to test this hypothesis.

(AEB)

16 An athlete returns the following times, in seconds, for 200 metres in fifteen races:

23·0 22·4 23·5 22·6 22·0 24·1 23·7 23·1 22·2 23·2 22·4 22·5 24·2 23·6 23·3

Assuming his times form a normal distribution and regarding the above as a random sample of 15 from this normal distribution, estimate the mean and variance.

Use these estimates to obtain the probability that, on a given occasion, the athlete will run 200 metres in less than 22·05 seconds.

Comment on the validity of this procedure.   (AEB)

17   A random sample of 18 students from Glasgow had a mean intelligence quotient of 109 with a standard deviation of 9. A random sample of 15 students from Edinburgh had a mean intelligence quotient of 113 with a standard deviation of 10. Is there a significant difference between the students as shown by these samples (i) at the 1% and (ii) at the 5% level of significance?   (AEB)

18   Ten wooden panels are painted partly with an ordinary paint and partly with a flame-resisting paint. The times, in suitable units, taken to burn through the panels are as follows:

Panel	A	B	C	D	E	F	G	H	I	J
Ordinary paint	12	10	8	20	6	13	7	9	25	15
Flame resisting	17	16	15	22	13	20	13	18	24	18

Examine the hypothesis that the paint does not make any real difference, stating any assumptions you make.   (AEB)

19   Eight typists were given electric typewriters instead of their previous machines, and the increases in the numbers of finished quarto pages in a day were found to be as follows:

$$8 \quad 12 \quad -3 \quad 4 \quad 5 \quad 13 \quad 11 \quad 0$$

Do these figures indicate that the typists in general work faster with electric typewriters? State the assumptions you make.   (AEB)

20   The heights in centimetres, of small random samples of boys and girls of age thirteen were as follows:

Boys	150	145	147·5	155	160	142·5	140	
Girls	142·5	147·5	130	127·5	147·5	150	142·5	140

Carry out a sample *t*-test to test whether the mean heights of the boys and girls differ at the 5% level of significance.

If you were asked to test whether the mean height of the boys differed from 150 cm, would you use a standard error estimate based only on the boys' heights or, as in the test just performed, one based on the heights of all children? Justify your answer.   (C)

CHAPTER 17

The F-distribution

Purpose of the F-test

When applying the t-test to the means of samples we assumed there was a common parent variance. If the sample variances differ so much that we cannot assume a common parent variance the t-test is nullified. The following test is used to see if two estimated population variances differ significantly, that is to test if they come from different parent populations.

If two samples of size n_1 and n_2 come from the same normal population we have two estimates of the variance of the parent population. Let $\sigma_{e_1}^2$ and $\sigma_{e_2}^2$ be the two estimated variances then $\sigma_{e_1}^2/\sigma_{e_2}^2$ should approach unity for large values of n_1 and n_2. In practice the ratio of these estimates may differ widely from unity, but significant points can be calculated for the ratio, and if for a particular pair of samples the selected percentage point is exceeded we conclude the samples are not from the same parent population.

The Distribution

The variance ratio is denoted by F, and tables have been giving the significant points at different levels of probability. F is a function of the number of degrees of freedom v_1, v_2 used in obtaining the estimated variances. Thus

$$v_1 = n_1 - 1 \quad \text{and} \quad v_2 = n_2 - 1, \quad \text{and} \quad F = \sigma_{e_1}^2/\sigma_{e_2}^2$$

The distribution for F is not symmetrical and it is convenient in practice to calculate the ratio F with the larger variance in the numerator so that $F_{v_2}^{v_1}$ is always greater than one, and v_1 is the number of degrees of freedom for the greater variance. This allows for a simpler tabulation of probabilities.

253

EXAMPLE

Test whether the following two samples come from the same parent population.

Sample A	37	36	37	38	37					
Sample B	38	39	38	37	36	39	39	39	38	37

SOLUTION

	Sample A			Sample B	
x_1	$(x_1 - m_1)$	$(x_1 - m_2)^2$	x_2	$(x_2 - m_2)$	$(x_2 - m_2)^2$
37	0	0	38	0	0
36	-1	1	39	1	1
37	0	0	38	0	0
38	1	1	37	-1	1
37	0	0	36	-2	4
5 185	$n_1 - 1 = 4$	2	39	1	1
$m_1 = 37$	$o_{e_1}^2 = 0{\cdot}5$		39	1	1
			39	1	1
			38	0	0
			37	-1	1
			10 380	$n_2 - 1 = 9$	10
			$m_2 = 38$	$o_{e_2}^2$	$1{\cdot}1111$

Therefore
$$F_4^9 = \frac{1{\cdot}1111}{0{\cdot}5} = 2{\cdot}2222$$

From the F-table we see that for $v_1 = 9$ and $v_2 = 4$ the 5% significant point for F lies between 6·04 and 5·96 and our value of F is much less than 5·96 and conclude our samples probably come from the same parent universe.

Interpolation

In the example it was not necessary to interpolate for the value of F but if the necessity does occur it is better to interpolate linearly on the *reciprocal* of the number of degrees of freedom. Thus $F_4^8 = 6{\cdot}04$; $F_4^{10} = 5{\cdot}96$.

Therefore
$$F_4^9 = 6{\cdot}04 - \frac{(1/8)-(1/9)}{(1/8)-(1/10)} \times (6{\cdot}04 - 5{\cdot}96) = 5{\cdot}996 = 6{\cdot}00$$

Since both variances probably refer to the same parent population the best estimate of the population variance is by formula (16.1). That is

$$\frac{0{\cdot}5 \times 4 + 1{\cdot}1111 \times 9}{4+9} = 0{\cdot}923$$

The z-Statistic

Because F cannot be accurately linearly interpolated sometimes the statistic z is used where

$$z = \tfrac{1}{2}\log_e F$$
$$= \log_e \sigma_{e_1} - \log_e \sigma_{e_2}$$

and the linear interpolation of this statistic is more accurate. Tables for the significant values of z are published and in our example

$$z = \tfrac{1}{2}\log_e 2\cdot222 = 0\cdot3992$$

From z-tables for $v_1 = 9$ and $v_2 = 4$ we have the 5% significant point for z is $0\cdot891$ and since z is less than this we verify the first solution.

EXAMPLE

Two pupils make repeated careful measurements of the same leaf. Their results in centimetres are:

Pupil A	2·40	2·29	2·37	2·41	2·48	2·35		
Pupil B	2·38	2·37	2·40	2·33	2·32	2·39	2·40	2·34

Do these estimates establish the fact that B is the more precise measurer?

SOLUTION

Calculate the two estimated variances.

Pupil A

$\Sigma x_1 = 14\cdot20$; $m_1 = 14\cdot20/6 = 2\cdot37$

$(x_1 - m_1)$	3	−8	0	4	1	−2	(Hundredths of cm)
$(x_1 - m_1)^2$	9	64	0	16	1	4	$\Sigma(x_1 - m_1)^2 = 94$

$$\sigma_{e_1}^2 = \frac{94}{5} = 18\cdot8$$

Pupil B

$\Sigma x_2 = 18\cdot93$; $m_2 = 18\cdot93/8 = 2\cdot37$

$(x_2 - m_2)$	1	0	3	−4	−5	2	3	−3	
$(x_2 - m_2)^2$	1	0	9	16	25	4	9	9	$\Sigma(x_2 - m_2)^2 = 73$

$$\sigma_{e_2}^2 = \frac{73}{7} = 10\cdot4$$

$$F_7^5 = \frac{\sigma_{e_1}^2}{\sigma_{e_2}^2} = \frac{18\cdot8}{10\cdot4} = 1\cdot8$$

With $v_1 = 5$ and $v_2 = 7$ the 5% significant value is 3·97. This means the smaller deviations of B's readings could be the result of random sampling and there is no evidence to conclude that B is a more precise measurer.

EXAMPLE

A sample of size 14 gives an estimated population variance of 3·2. Could this sample have come from a large population of variance 9·6?

SOLUTION

This is equivalent to comparing two estimated variances one with an infinite number of degrees of freedom and variance 9·6, and the other of 13 degrees of freedom and variance 3·2.

$$F^{\infty}_{13} = \frac{9·6}{3·2} = 3$$

For $v_1 = \infty$ and $v_2 = 13$ the 5% significant value is 2·21, the $2\frac{1}{2}$% is 2·60, and the 1% is 3·17. In our problem the significant value lies between $2\frac{1}{2}$% and 1%. Therefore it is very highly probable that the sample is not derived from the specified population.

Exercise 17.1

1   Find the 5% significant value for (i) F^{12}_9, (ii) F^{24}_9, (iii) F^{18}_9.

2   A sample of nine measurements has an estimated population variance of 2·63 and a sample of seventeen measurements has an estimated population variance of 5·02. Can these samples be reasonably regarded as coming from the same normal population?

3   To test if practice improves the ability to measure, a worker made 14 measurements of the same object and after considerable practice a further 10 measurements. The results were:

First test (cm)	4·50	4·53	4·90	4·69	4·72	4·39	4·77	4·45	4·58	4·59
	4·61	4·42	4·56	4·40						
Second test	4·52	4·64	4·71	4·60	4·53	4·55	4·47	4·72	4·42	4·70

Is there evidence to suggest that practice has improved the worker's ability to measure?

4   The heights of two random samples of pupils aged sixteen from two different schools in the same borough are:

First school
Height (cm)   157·5   162·5   170   172·5   177·5   180

Second school
Height (cm)   152·5   157·5   162·5   167·5   172·5   172·5   175   177·5   180   182·5

Is there evidence that the pupils in one school are taller than the pupils in the other school?

5   To test if one operator is a more consistent worker than another they were asked to make components using the same machine so that variability can be assumed to be a function of the operator.

First operator
Length of component (cm)   5·02   5·06   5·10   5·04   5·03   5·04
                             5·09   5·03   5·04   5·05   5·08   5·02

Second operator
Length of component (cm)   5·07   5·09   5·17   5·07   5·14   5·05
                             5·07   5·13   5·12   5·12   5·16   5·13

Is there evidence to show that one operator is more consistent than the other?

6   Two operators are tested with the same machine for precision in making measurements of an article. One operator makes 25 readings with a sample standard deviation of 1·43, and the other 37 readings with a standard deviation of 0·99. Is the difference in the deviations significant at the (i) 5%, (ii) 1% level?

7   The voltages of ten dry cells of type A are found to be 1·50, 1·46, 1·52, 1·52, 1·47, 1·52, 1·49, 1·45, 1·51, 1·49 and those of eight dry cells of type B are 1·48, 1·46, 1·57, 1·54, 1·59, 1·46, 1·56, 1·58. Is there reason to believe that the distributions of voltages in the two types of batteries are different and, if so, in what way? (AEB)

8   An explorer measured the lengths of the feet of ten adult males of tribe A with the following results: 23·9, 27·4, 27·7, 22·9, 25·4, 28·2, 26·7, 24·1, 24·6, 24·9 centimetres.

He also measured the lengths of the feet of twelve men of tribe B, obtaining: 21·3, 22·4, 25·4, 26·2, 23·6, 21·3, 22·9, 21·6, 21·6, 23·1, 21·1 21·8 centimetres.

Is there evidence of a real difference between the distributions of lengths of feet in the two tribes? (AEB)

9   Packets of detergent sell at the same price.

Ten packets of Ono were bought, but only seven packets of Draff could be obtained. The masses of detergent in the packets of the two types were found to be, in grammes:

Ono: 243·8, 246·9, 241·5, 252·6, 231·6, 250·3, 250·6, 245·2, 249·9, 243·5
Draff: 251·2, 252·6, 250·9, 252·3, 253·2, 254·9, 246·9

Is there reason to suppose that
 (i) the amount in packets of Draff is more consistent than in packets of Ono,
(ii) the average amount in packets of Draff is greater than the average weight in packets of Ono?   (AEB)

CHAPTER 18

Quality Control, Range and Standard Deviation

Quality Control

Even if there is no wear on a machine any product is liable to variation and this variation is assumed to be normally distributed. A normal distribution is completely specified if its mean and standard deviation are known and these correspond to the setting and precision, respectively, of the machine. If these properties of the machine are known the number of articles that fall outside any given limits can be calculated by means of the property of the normal curve.

It is the purpose of Quality Control to obtain warning of the development of any change in the setting or precision of a machine by comparing the proportions of observed values falling within chosen ranges of the mean and standard deviation and those expected on the assumption that the distribution of variability is normally distributed. For this purpose inner and outer control limits are chosen which are sometimes called warning and action limits, respectively.

Thus, on the normal hypothesis, one in forty of observed values would exceed $\mu + 1.96\sigma$ and one in forty would be less than $\mu - 1.96\sigma$, while one in a thousand observed values would exceed $\mu + 3.09\sigma$ and one in a thousand would be less than $\mu - 3.09\sigma$.

Quality Control Schemes

When producing a Quality Control Scheme for any machine the first step is to estimate the mean and standard deviation of the measurement to be controlled under stable conditions of production. To do this, about one hundred measurements are taken and the mean and standard deviation calculated. The mean is easily calculated, but the calculation of the standard deviation can be laborious, and it is useful if a fairly accurate estimate can be made by simpler means.

Standard Deviation and Range

Fortunately there is a simple method of estimating the standard deviation from the range of samples of size n from a normal population, and this has the double advantage of easy calculation in the subsequent sampling, and thus allowing comparatively inexpert operators to keep a check on the process, and of giving an estimate of σ independent of any change of mean which occurs without a change of variance. The sampling distribution of range is one of the more difficult to derive mathematically, but data sufficient for the purposes of Quality Control have been tabulated.

If x_1 and x_n are the lowest and highest values of the variate in a sample of n observations and ω_n stands for the range in the sample, then

$$\omega_n = x_n - x_1$$

If a_n is an appropriate numerical factor it can be shown that $a_n\omega_n$ is an unbiased estimate of the population standard deviation. In some tables the reciprocal of a_n has been calculated and is written as d_n, that is, in this case, an unbiased estimate of the parent standard deviation is ω_n/d_n (see Table 7 for the values of a_n).

Accuracy of the Estimate of σ

In samples of size 10 or less items the range estimate of σ is as good as the estimate obtained by any other method, but as the sample size increases the accuracy of the estimate by means of the range falls off progressively. For this reason it is common in practice to estimate σ from the mean range of the observations in a number of small groups rather than with individual readings. Thus, if a number of small samples of size n observations are made and the mean value of their ranges is denoted by $\bar{\omega}_n$ then $a_n\bar{\omega}_n$ may be used as an estimate of σ.

EXAMPLE

The following samples are from a normal population. Estimate the population standard deviation by (i) using the mean range of the samples, (ii) by calculation by the usual method of the standard deviation of the 50 items.

				Sample					
1	2	3	4	5	6	7	8	9	10
10·18	11·70	10·48	11·85	12·39	12·57	10·84	11·38	11·37	11·07
12·37	12·03	8·48	10·25	10·10	9·63	12·29	9·11	12·30	10·26
10·13	11·63	12·07	10·82	12·01	11·47	11·52	10·33	10·07	13·84
12·13	9·50	11·30	11·01	11·61	10·10	9·38	12·19	11·26	9·89
10·80	11·13	11·11	12·54	10·72	11·42	12·24	9·95	11·38	11·16

SOLUTION

(i) ω_n for each sample is

| 2·24 | 2·53 | 3·59 | 2·29 | 2·29 | 2·94 | 2·91 | 3·08 | 2·23 | 3·95 |

and $\bar{\omega}_n$ equals

$$\frac{\Sigma \omega_n}{10} = \frac{28·05}{10} = 2·805$$

From Table 7 the value of a_n corresponding to $n = 5$ is 0·4299. Therefore the estimate of the population σ is

$$a_n \bar{\omega}_n = 2·805 \times 0·4299 = 1·21$$

(ii) The simplest way to calculate the standard deviation of the 50 items is to calculate Σx and Σx^2 and use the formula

$$\sigma^2 = \frac{\Sigma x^2}{N} - \left(\frac{\Sigma x}{N}\right)^2$$

The sum of the fifty items is 555·36. To find Σx^2 square each item and sum

					Sample				
1	2	3	4	5	6	7	8	9	10
103·7	136·9	109·9	140·3	153·5	158·0	117·4	129·5	129·3	122·6
153·0	144·7	71·9	105·0	102·0	92·7	151·0	83·0	151·3	105·3
102·6	135·3	145·7	117·0	144·2	131·6	132·8	106·7	101·5	191·5
147·1	90·3	127·7	121·2	134·8	102·0	88·0	148·6	126·8	97·8
116·6	123·9	123·4	157·3	114·9	130·5	149·8	99·0	129·5	124·6

Therefore $\Sigma x^2 = 6223·7$, and

$$\sigma_e^2 = \frac{6223·7}{50} - \left(\frac{555·36}{50}\right)^2$$

$$= 1·50$$

Therefore, $\qquad \sigma_e = 1·23$

Thus the estimated value of σ obtained from the range of the samples agrees very closely with the best estimate of σ calculated from the fifty items.

Control Limits for Individual Items

If the parent population is normally distributed we can set up control limits to test whether a machine is producing articles to specification.

Thus, if in the last example we want to find the limits which 1/40 of the articles may lie outside we know the limits are $M \pm 1.96\sigma$.

Now $\qquad M = 555.36/50 = 11.11$, and $\sigma_e = 1.21$

Therefore, $M \pm 1.96\sigma = 11.11 \pm 1.96 \times 1.21 = 13.5$ and 8.7

The limits outside which 1/1000 of the items may lie are

$$M \pm 3.09\sigma = 11.1 \pm 3.09 \times 1.21 = 14.8 \quad \text{and} \quad 7.4$$

Control Charts

It is usual to draw a chart giving the number of the item and the control limits marked. Selected items are then measured and the result plotted on the chart. Thus, for our 50 items we have Diagram 18.1.

It will be noticed that two of the items fall outside the inner control lines which gives a warning that the machine may need checking.

Control Limits for the Means of Samples

If the parent population is not quite normally distributed it is known that the means of samples of size n may be assumed to be normally distributed with the same mean μ as the parent population and with standard deviation $\sigma/\sqrt{n}$. The inner and outer control limits for the means of samples are $M \pm (1.96\sigma/\sqrt{n})$ and $M \pm (3.09\sigma/\sqrt{n})$. If σ is not given the estimate for σ is $a_n \bar{\omega}_n$ and the estimate for the standard deviation of the means of samples of size n is $a_n \bar{\omega}_n/\sqrt{n}$ and the control limits for the means of samples of size n will be

$$M \pm \frac{1.96 a_n \bar{\omega}_n}{\sqrt{n}} \quad \text{and} \quad M \pm \frac{3.09 a_n \bar{\omega}_n}{\sqrt{n}}$$

where M is the mean of the initial set of observations.

For convenience the quantities $1.96 a_n/\sqrt{n}$ and $3.09 a_n/\sqrt{n}$ can be calculated for any given value of n and any other control limits. Denoting $1.96 a_n/\sqrt{n}$ and $3.09 a_n/\sqrt{n}$ by $A_{0.025}$ and $A_{0.001}$, respectively, we have the following table.

n	2	3	4	5	6	7	8	9	10
$A_{0.025}$	1.23	0.67	0.48	0.37	0.32	0.27	0.24	0.22	0.20
$A_{0.001}$	1.94	1.05	0.75	0.59	0.50	0.43	0.38	0.35	0.32

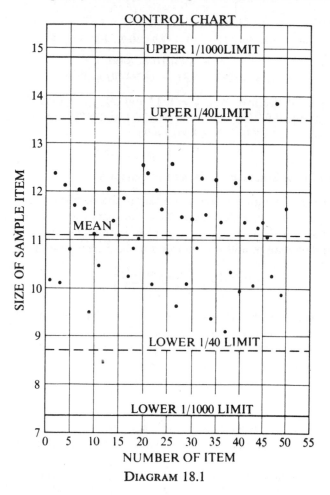

DIAGRAM 18.1

EXAMPLE

The length of a leaf is normally distributed. The following 10 samples are of five randomly selected leaves. Calculate 1/40 and 1/1000 limits for the mean length of leaves in samples of size five.

Samples

1	2	3	4	5	6	7	8	9	10
10·8	12·2	13·7	13·4	13·1	3·4	12·5	4·9	8·8	8·3
6·1	11·1	8·7	17·0	13·6	6·1	11·0	9·0	9·3	6·4
11·5	8·2	13·0	10·3	7·0	7·1	7·1	11·4	13·9	8·7
7·5	6·8	2·2	13·8	9·5	12·1	11·9	11·2	12·3	13·5
5·8	12·0	6·4	6·9	5·7	14·3	5·2	13·9	10·2	8·6

SOLUTION

$$\omega \qquad 5\cdot7 \quad 5\cdot4 \quad 11\cdot5 \quad 10\cdot1 \quad 7\cdot9 \quad 10\cdot9 \quad 7\cdot3 \quad 9\cdot0 \quad 5\cdot1 \quad 7\cdot1$$

Therefore,
$$\bar{\omega} = 80\cdot0/10 = 8\cdot00$$
$$\Sigma x = 487\cdot4 \qquad M = 487\cdot4/50 = 9\cdot748$$

For samples of size five, $A_{0\cdot025} = 0\cdot37$ and $A_{0\cdot001} = 0\cdot59$. Therefore the inner limits equal $9\cdot748 \pm 0\cdot37 \times 8\cdot0 = 12\cdot71$ and $6\cdot79$, and the outer limits equal $9\cdot748 \pm 0\cdot59 \times 8\cdot0 = 14\cdot52$ and $4\cdot98$.

Variation in the Standard Deviations of Samples

If we wish to maintain a certain precision, the samples must be tested to see that the standard deviation does not vary beyond certain limits. If samples of size n are taken from an infinite normal population of standard deviation σ, and the sample standard deviations s, where

$$s = \sqrt{\left[\frac{\Sigma(x - \bar{x}^2)}{n} \right]}$$

and arranged in a frequency table, it is found that for small samples the distribution is slightly positively skew, but it approaches the normal distribution with an increase in n. So for small samples the means of the standard deviations are not at σ. The mean value of the standard deviations of samples of size n, (μ_s), is given by

$$\mu_s = c_n \sigma$$

where c_n is a factor depending on the magnitude of n. The values of c_n can be calculated for various values of n. They are given in the following table.

n	2	3	4	5	6	7
c_n	0·5642	0·7236	0·7979	0·8407	0·8686	0·8882
	8	9	10	11	12	13
	0·9027	0·9139	0·9228	0·9300	0·9359	0·9410
	14	15	16	17	18	19
	0·9453	0·9490	0·9523	0·9551	0·9577	0·9599
	20	21	22	23	24	25
	0·9619	0·9638	0·9655	0·9670	0·9684	0·9697

For values of n greater than 25 the following approximation

$$c_n \simeq \left(1 - \frac{3}{4n} \right)$$

is accurate enough.

This relation gives us an unbiased estimate of σ. For if

$$s = c_n \sigma$$

then $$\sigma = s/c_n$$

If the parent population is approximately normal the standard deviation of the distribution of s is $\sigma/\sqrt{(2n)}$. Now we have enough information to set up inner and outer control limits for the values of s. The mean value of s is $c_n\sigma$, so the inner and outer control limits are

$$c_n\sigma \pm \frac{1\cdot96\sigma}{\sqrt{(2n)}} \quad \text{and} \quad c_n\sigma \pm \frac{3\cdot09\sigma}{\sqrt{(2n)}}$$

or

$$\left[c_n \pm \frac{1\cdot96}{\sqrt{(2n)}} \right]\sigma \quad \text{and} \quad \left[c_n \pm \frac{3\cdot09}{\sqrt{(2n)}} \right]\sigma$$

EXAMPLE

A manufacturer wants a product made to the specifications $\mu = 35$ cm; $\sigma = 0\cdot5$ cm. Determine the 1/40 and 1/1000 control limits for quality control for the means and standard deviations of samples of size 10.

SOLUTION

Means.
$$\text{Inner control limits} = M \pm \frac{1\cdot96\sigma}{\sqrt{n}}$$

$$= 35 \pm \frac{1\cdot96 \times 0\cdot5}{\sqrt{10}}$$

$$= 35\cdot30 \text{ and } 34\cdot70$$

$$\text{Outer control limits} = M \pm \frac{3\cdot09\sigma}{\sqrt{n}}$$

$$= 35 \pm \frac{3\cdot09 \times 0\cdot5}{10}$$

$$= 35\cdot49 \text{ and } 34\cdot51$$

Standard deviation. $n = 10$. Therefore, $c_n = 0\cdot9228$

$$\text{Inner control limits} = \left[c_n \pm \frac{1.96}{\sqrt{(2n)}}\right]\sigma$$

$$= \left(0.9228 \pm \frac{1.96}{\sqrt{20}}\right)0.5$$

$$= 0.68 \text{ and } 0.24$$

$$\text{Outer control limits} = \left[c_n \pm \frac{3.09}{\sqrt{(2n)}}\right]\sigma$$

$$= \left(0.9228 \pm \frac{3.09}{\sqrt{20}}\right)0.5$$

$$= 0.81 \text{ and } 0.12$$

EXAMPLE

The length of a leaf is normally distributed. The following 10 samples are of five randomly selected leaves. Calculate 1/40 and 1/1000 control limits for (i) the mean length of the leaves, (ii) the precision of the measurement.

				Samples					
1	2	3	4	5	6	7	8	9	10
10·8	12·2	13·7	13·4	13·1	3·4	12·5	4·9	8·8	8·3
6·1	11·1	8·7	17·0	13·6	6·1	11·0	9·0	9·3	6·4
11·5	8·2	13·0	10·3	7·0	7·1	7·1	11·4	13·9	8·7
7·5	6·8	2·2	13·8	9·5	12·1	11·9	11·2	12·3	13·5
5·8	12·0	6·4	6·9	5·7	14·3	5·2	13·9	10·2	8·6

SOLUTION

	1	2	3	4	5	6	7	8	9	10
$\omega =$	5·7	5·4	11·5	10·1	7·9	10·9	7·3	9·0	5·1	7·1
$M =$	8·34	10·06	8·8	12·28	9·78	8·6	9·54	10·08	10·9	9·1
$s = a_n\omega =$	2·45	2·32	4·94	4·34	3·40	4·69	3·14	3·14	2·19	4·05

$\omega = $ range = (in sample 1), $11.5 - 5.8 = 5.7$

$M = $ mean of sample

$s = $ standard deviation of the sample. Thus, in sample 1, $s = a_n\omega = 0.4299 \times 5.7 = 2.45$

$\bar{\omega} = 80.0/10 = 8.00 \qquad \Sigma x = 487.4$

mean of the 50 leaves $= 487.4/50 = 9.748$

For samples of size five, $A_{0.025} = 0.32$, and $A_{0.001} = 0.50$. Therefore, the inner limits equal $9.748 \pm 0.32 \times 8.0 = 12.30$ and 7.19, and the outer limits equal $9.748 \pm 0.50 \times 8.0 = 13.75$ and 5.75. It will be noticed that all the sample means are within the inner control limits.

Control limits for the standard deviations. $c_n = 0.8407$.

$$\text{Inner control limits} = \left[c_n \pm \frac{1.96}{\sqrt{(2n)}} \right] \sigma$$

$$= \left(0.8407 \pm \frac{1.96}{\sqrt{10}} \right) 3.44$$

$$= 5.02 \text{ and } 0.76$$

$$\text{Outer control limits} = \left[c_n \pm \frac{3.09}{\sqrt{(2n)}} \right] \sigma$$

$$= \left(0.8407 \pm \frac{3.09}{\sqrt{10}} \right) 3.44$$

$$= 6.20 \text{ and } -0.52$$

Since s cannot be less than zero the outer control limits are 6.20 and 0. The sample standard deviations are all within the inner control limits.

Exercise 18.1

1 The following sets of samples are from normally distributed populations. Estimate

(a) the population mean,
(b) the population standard deviation by
    (i) using the mean range of the samples,
    (ii) the formula $\quad \sigma^2 = \dfrac{\Sigma x^2}{n} - \left(\dfrac{\Sigma x}{n} \right)^2$

(c) Calculate 1/40 and 1/1000 limits for the distribution of sample means and sample standard deviations.

(1)

				Samples					
1	2	3	4	5	6	7	8	9	10
8·8	4·3	2·0	2·7	3·7	7·4	0·3	8·4	5·6	3·6
6·4	2·4	3·0	10·9	7·5	5·2	4·2	7·4	5·0	8·5
11·5	4·2	5·0	3·5	6·6	2·3	3·8	9·5	6·1	3·7
6·1	6·6	8·3	6·8	5·4	4·2	4·5	5·1	4·1	1·4
8·2	8·8	9·8	2·6	2·3	4·1	5·2	6·8	3·3	3·3

(2)

				Samples					
1	2	3	4	5	6	7	8	9	10
4·8	−1·8	−1·0	9·0	0	−2·8	−3·4	2·4	−9·6	5·0
0·4	−9·4	0·4	0·2	2·2	7·0	0·4	2·8	3·8	8·6
−5·4	−1·6	6·8	3·6	−1·8	−2·6	4·6	−2·0	−5·8	4·2
−1·6	−2·4	4·8	1·2	−3·4	−7·2	7·8	−10·1	2·0	−5·8

2   Explain what is meant by an unbiased estimate. Show that for a random sample x_i (i = 1, 2, 3, . . ., n) of n observations, with mean $\bar{x}$, an unbiased estimate of the population variance is

$$\frac{1}{n-1} \Sigma(x_i - \bar{x})^2$$

Rods are being made with a nominal diameter of 5·64 cm. A random sample of ten is measured and found to have diameters 5·613, 5·682, 5·636, 5·671, 5·652, 5·629, 5·677, 5·645, 5·639, 5·668. Estimate the variance of the population (i) by the above formula, and (ii) using the range. Find 95% confidence limits for the population mean.   (AEB)

3   Eight trees of a rare species are found to have girths 76·5, 84·4, 69·2, 75·0, 62·2, 68·0, 70·4, 69·8 cm. Estimate the mean and variance of the girth of this species of tree, and give 95% confidence limits for the mean by using

(i) the unbiased estimate of the population variance
(ii) the range of the sample to estimate the standard deviation of the population.   (AEB)

4   Cylindrical rods are to be manufactured with a diameter between 0·635 ± 0·0076 cm, and a length between 7·62 ± 0·0254 cm. It is found that in the ordinary course of production the diameters follow a normal distribution with standard deviation 0·00254 cm. The lengths form a normal distribution with mean 7·62 cm and standard deviation 0·00102 cm.

Describe, with diagrams, the control charts you would set up in order to check production.   (AEB)

5   In an industrial manufacturing process, when production is under control, one unit in fifty is defective. A sample of 100 units is taken at random every half-hour. Sketch a control chart for the number of defectives found in a sample, showing the 1 in 40 and 1 in 1000 lines.

Something goes wrong with the process, so that one unit in twenty is defective. Calculate the probability that the number in a random sample of 100 will go outside (i) the 1 in 40 line, (ii) the 1 in 1000 line.

What is the probability that the process will continue out of control for two hours without any sample exceeding the 1 in 1000 limit?   (AEB)

CHAPTER 19

Significance of Sample Correlation Coefficients

Problems connected with Sample Correlation Coefficients

The problems connected with sample correlation coefficients are similar to all sampling problems.

   (i) Does the sample correlation coefficient give evidence of correlation in the parent distribution?

  (ii) Does the sample correlation coefficient differ significantly from a given value?

 (iii) Do two sample correlation coefficients give evidence that the samples are from the same parent population?

 (iv) Given several sample correlation coefficients how do we find the best average correlation coefficient?

The sample correlation coefficient is denoted by r and the parent population coefficient by ρ. Unfortunately the distribution of sample correlation coefficients is not normal and so a statistic must be computed for each of the problems. A population consisting of two paired variables is known as a bivariate distribution and in the following discussion we assume the distribution of each variable to be normal.

PROBLEM (i)

To test if r differs significantly from zero. If it is assumed $\rho = 0$ then the statistic

$$r \times \sqrt{\left(\frac{n-2}{1-r^2}\right)}$$

has the t-distribution about the mean ρ with $n-2$ degrees of freedom

EXAMPLE

A random sample of 20 pairs of observations from a normal bivariate population has a correlation coefficient of 0·46. Is this value significant?

SOLUTION

For $r = 0.46$: $n = 20$, we have

$$t = 0.46 \times \sqrt{\left[\frac{20-2}{1-(0.46)^2}\right]} = 2.479$$

From the table of the t-distribution we see that for 15 and 20 degrees of freedom the 5% significant values are 2·13 and 2·09, respectively. Therefore our value of t, with 18 degrees of freedom is greater than these and is significant at the 5% value, and therefore the hypothesis of there being no correlation in the parent population is rejected.

EXAMPLE

Suppose the number of observations in the previous problem had been 10.

SOLUTION

Now

$$t = 0.46 \times \sqrt{\left[\frac{10-2}{1-(0.46)^2}\right]} = 1.652$$

For $v = 8$, P lies between 25% and 10% and t is not significant at the 5% level, and therefore the hypothesis that there is no correlation in the parent population is probably true.

PROBLEM (ii)

To test if r differs significantly from a given value. In this problem it is found that the statistic

$$z = \tfrac{1}{2}\log_e \frac{1+r}{1-r}$$

is approximately normally distributed with mean

$$\zeta = \tfrac{1}{2}\log_e \frac{1+\rho}{1-\rho}$$

and standard deviation $1/\sqrt{(n-3)}$.

The logarithms are taken to the base e but these can be converted into the base 10, for

$$\log_e x = 2.3026 \log_{10} x$$

and therefore
$$z = \tfrac{1}{2}\log_e \frac{1+r}{1-r}$$

$$= 1{\cdot}1513 \log_{10} \frac{1+r}{1-r}$$

and
$$\zeta = 1{\cdot}1513 \log_{10} \frac{1+\rho}{1-\rho}$$

Students who are familiar with the hyperbolic functions will recognize that

$$z = \tfrac{1}{2}\log_e \frac{1+r}{1-r} = \tanh^{-1}r$$

and tables giving the corresponding values of z and r are available and this is the easiest way of finding z. Conversely the tables can be used in a reciprocal manner and r can be read from the tables if z is known (see Table 4, p. 337).

EXAMPLE

A correlation coefficient of 0·70 is obtained from a sample of 30 observations. Is this value consistent with the assumption that the parent population coefficient is 0·80?

SOLUTION

From the tables on page 337 for $r = 0{\cdot}70$, $z = 0{\cdot}867$, and for ρ (i.e. r in Table 4) $= 0{\cdot}80$, ζ (i.e. z in the table) $= 1{\cdot}099$. The standard error or $\sigma = 1/\sqrt{(30-3)} = 0{\cdot}193$. Now $(z-\zeta)/\sigma$ is a normal deviate and equals

$$\frac{0{\cdot}867-1{\cdot}099}{0{\cdot}193} = -1{\cdot}20$$

From the normal tables we see this value is only significant at the 11·5% level, and we conclude that ρ may well be 0·80.

PROBLEM (iii)

To compare sample correlation coefficients. To test the significance of two correlation coefficients from two separate samples, we compare the difference of the two corresponding values of z with the standard error of the difference. Remember the variance of the difference of two statistics is the sum of their variances. The difference between the two values of z is normally distributed.

EXAMPLE

Two samples of sizes 30 and 40 pairs of observations have correlation coefficients of magnitude -0.3 and -0.4, respectively. Do these coefficients differ significantly at the 5% level?

SOLUTION

For $r_1 = -0.3$, $z_1 = -0.310$ and $\sigma_1^2 = 1/(30-3) = 0.037$.
For $r_2 = -0.4$, $z_2 = -0.424$ and $\sigma_2^2 = 1/(40-3) = 0.027$.
Therefore the variance of $z_1 - z_2 = 0.037 + 0.027 = 0.064$, and $\sigma = 0.25$.
$\dfrac{z_1 - z_2}{\sigma}$ is a normal variate and equals $\dfrac{-0.310 + 0.424}{0.25} = 0.45$

There is no significant difference and it is highly probable the two samples come from the same parent universe.

Note This test could be used to see if r differed significantly from $\rho = 0$, but it is not as reliable as the test quoted in Problem (i).

PROBLEM (iv)

To find the average correlation coefficient. If we have samples of $n_1, n_2, \ldots$ pairs of observations with correlation coefficients $r_1, r_2, \ldots$, respectively, then the mean value of r is given by

$$\bar{r} = \tanh \bar{z}$$

where $\bar{z}$ is the weighted mean of z and is given by

$$\bar{z} = \frac{v_1 z_1 + v_2 z_2 + \ldots}{v_1 + v_2 + \ldots}$$

and the standard error of $\bar{r}$ is

$$\frac{1}{\sqrt{(v_1 + v_2 + \ldots)}}$$

where $v_k = n_k - 3$. We must use a weighted mean because values calculated from larger samples are more accurate and therefore more important than values from smaller samples.

EXAMPLE

Four samples of 41, 36, 39, 44 pairs of observations drawn from the same normal bivariate population have correlation coefficients of 0.39, 0.43,

0·33, 0·36, respectively. Calculate the average value of the correlation coefficient and its standard error.

SOLUTION

It is most convenient to tabulate the work.

r	n	z	v	vz
0·39	41	0·412	38	15·66
0·43	36	0·460	33	15·18
0·33	39	0·343	36	12·35
0·36	44	0·377	41	15·46
			148	58·65

$$\bar{z} = \frac{58·65}{148} = 0·396$$

Therefore, $\qquad \bar{r} = 0·38$

and the standard error of

$$\bar{r} = \frac{1}{\sqrt{(38+33+36+41)}} = 0·082$$

FURTHER EXAMPLES

1   What is the least value of r in a sample of 22 pairs of observations that is significant at the 5% level?

SOLUTION

$$v = 20 \qquad t_{0.05} = 2·09$$

If the value of r is to be significant

$$\frac{r\sqrt{20}}{\sqrt{(1-r^2)}} \geqslant 2·09$$

Therefore, squaring and reducing

$$20r^2 \geqslant 4·368 - 4·368r^2$$
or $\qquad r \geqslant 0·423$

EXAMPLE 2

A value of r of 0·65 is calculated from a random sample of 36 pairs of observations from a normal bivariate population. Is this value of r consistent with the assumption that $\rho = 0.45$?

SOLUTION

If $r = 0.65$, $z = 0.775$, and if $\rho = 0.45$, $\zeta = 0.485$. Variance $= 1/(36-3)$ $= 0.0303$, and standard error $= 0.174$. So the normal variate

$$\frac{z-\zeta}{\sigma} = \frac{0.775-0.485}{0.174} = 1.67$$

This value is significant at the 5% level but not at the 1% level. Therefore there is some doubt as to the truth of the assumption.

EXAMPLE 3

Twenty-four students were given a test in mathematics and physics. The correlation coefficient between the two sets of marks was 0.75. What are the 95% confidence limits for ρ?

SOLUTION

$$\frac{|z-\zeta|}{\sigma} \leqslant 1.96$$

For $r = 0.75$, $z = 0.973$, and standard error

$$= \frac{1}{\sqrt{(24-3)}} = 0.2091$$

Therefore $|z-\zeta| \leqslant 1.96 \times 0.2091 \leqslant 0.4098$

that is $0.563 \leqslant \zeta \leqslant 1.383$

Therefore the minimum value of ρ is given by

$$\rho = \tanh 0.563 = 0.51$$

The maximum value of ρ is given by

$$\rho = \tanh 1.383 = 0.88$$

That is $0.51 \leqslant \rho \leqslant 0.88$

EXAMPLE 4

The correlation coefficients 0.52 and 0.33 are obtained from two random samples of size 29 and 36 pairs, respectively. Is there a significant difference between these coefficients at the 5% level?

SOLUTION

Since $r_1 = 0.52$, $z_1 = 0.576$, and variance$_1 = 1/(29-3) = 1/26$.
Since $r_2 = 0.33$, $z_2 = 0.343$, and variance$_2 = 1/(36-3) = 1/33$.

Therefore
$$\sigma_{z_1-z_2} = \sqrt{\left(\frac{1}{26}+\frac{1}{33}\right)} = 0.2622$$

Therefore on the assumption that the samples come from the same population the normal deviate of the difference of z_1 and z_2 is

$$\frac{(0.576-0.343)-0}{0.2662} = 0.877$$

Using a two-tailed test of the normal distribution we can only reject the hypothesis if the normal deviate is $\geqslant 1.96$ or $\leqslant -1.96$. This is not so and we conclude the two samples most likely come from the same parent population.

Exercise 19.1

1   A correlation coefficient based on a sample size of 38 was calculated as 0.4. Is there any evidence to show that the population correlation coefficient differs from zero at the (i) 5%, (ii) 1% level of significance?

2   Repeat question 1 with a sample size of 6.

3   A correlation coefficient from a sample of size 34 was calculated as 0.49. Is there evidence to assume the population correlation coefficient is (i) as small as 0.29, (ii) as big as 0.72 at the 5% level of significance?

4   Find (i) 95%, (ii) 99% confidence limits for the population correlation coefficient if a sample of size 28 gives a correlation coefficient of 0.61.

5   Correlation coefficients obtained from two samples of size 28 and 33 were calculated to be 0.75 and 0.80, respectively. Is there evidence that they come from the same parent population at the 5% level of significance?

6   Three sampling correlation coefficients from samples of size 19, 28, 39 pairs of observations are 0.67, 0.69 and 0.70, respectively. Calculate the average value of the correlation coefficients and its standard error.

7   The table shows the examination marks of 8 students in Algebra and Statistics. Calculate the product-moment coefficient of correlation between the Algebra and Statistics marks. Explain the significance of your answer.

Algebra	10	24	30	35	41	59	68	70
Statistics	30	47	44	71	60	89	99	74

(AEB)

Calculate the significant probability.

CHAPTER 20

Significance of Spearman's Rank Correlation Coefficient

Introduction

The advantage of Spearman's Rank Coefficient is that it does not require the assumption that the variables are normally distributed. The population coefficient is generally denoted by ρ and the sample coefficient by r_s. If n is the number of pairs of observations in a sample then

$$r_s = 1 - \frac{6\Sigma d^2}{n(n^2-1)}$$

where Σd^2 is the sum of the squares of the difference in the ranking of two corresponding observations.

Distribution of Σd^2

For samples of size n the variable in r_s is Σd^2. If the first variable is arranged in its natural ranking order 1, 2, 3, . . . n, then the second variable may be ranked in any of $n!$ ways. (We are assuming there are no ties). Thus if $n = 4$ there are 24 ways of ranking the second variable. For small values of n it is possible to calculate the value of Σd^2 for all these rankings and since different rankings can give us the same value of Σd^2 it is possible to form a probability distribution for Σd^2. For example if $n = 4$ we get $4! = 24$ possible rankings for the second variable. These rankings with the corresponding value of Σd^2 are given below.

Rankings

	1 2 3 4	1 2 3 4	1 2 3 4	1 2 3 4	1 2 3 4	1 2 3 4
	1 2 3 4	4 1 2 3	3 4 1 2	2 3 4 1	1 2 4 3	3 1 2 4
$\Sigma d^2 =$	0	12	16	12	2	6

	1 2 3 4	1 2 3 4	1 2 3 4	1 2 3 4	1 2 3 4	1 2 3 4
	4 3 1 2	2 4 3 1	1 3 4 2	2 1 3 4	4 2 1 3	3 4 2 1
$\Sigma d^2 =$	18	14	6	2	14	18

1 2 3 4	1 2 3 4	1 2 3 4	1 2 3 4	1 2 3 4	1 2 3 4
1 4 3 2	2 1 4 3	3 2 1 4	4 3 2 1	1 3 2 4	4 1 3 2

$\Sigma d^2 =$ 8   4   8   20   2   14

1 2 3 4	1 2 3 4	1 2 3 4	1 2 3 4	1 2 3 4	1 2 3 4
2 4 1 3	3 2 4 1	1 4 2 3	3 1 4 2	2 3 1 4	4 2 3 1

$\Sigma d^2 =$ 10   14   6   10   6   18

Arranging the values of Σd^2 as a frequency distribution we have:

Σd^2	0	2	4	6	8	10	12	14	16	18	20
f	1	3	1	4	2	2	2	4	1	3	1
Cum. freq.	24	23	20	19	15	13	11	9	4	4	1

It is now possible to calculate the probability of obtaining a given value of Σd^2 on random ranking. The probability of $\Sigma d^2 = 20$ is 1/24 or 0·04167, the probability of getting a value of Σd^2 equal to, or more than 18 is 4/24, or 0·16668, and so on.

These probabilities can be arranged as follows:

$\Sigma d^2 \geqslant$	0	2	4	6	8	10	12	14
P	1·000	0·958	0·833	0·792	0·625	0·542	0·458	0·375
	16	18	20					
	0·208	0·167	0·042					

If we wish to know the probability that Σd^2 is less than, or equal to a given value, all we have to do is reverse the table, for the distribution of Σd^2 is symmetrical. Thus,

$\Sigma d^2 \leqslant$	0	2	4	6	8	10	12	14
P	0·042	0·167	0·208	0·375	0·458	0·542	0·625	0·792
	16	18	20					
	0·833	0·958	1·000					

Let us examine what these tables tell us. If $n = 4$;

$$r_s = 1 - \frac{6 \times \Sigma d^2}{4(16-1)} = 1 - \frac{\Sigma d^2}{10}$$

and if $r_s = 0$ then $\Sigma d^2 = 10$. If $r_s = 1$ then $\Sigma d^2 = 0$, and if $r_s = -1$ then $\Sigma d^2 = 20$. The tables tell us that on random sampling the chances of obtaining a value of $\Sigma d^2 = 20$, or $r_s = -1$, is 0·042, and the chances of obtaining a value of $\Sigma d^2 = 0$, or $r_s = 1$ is also 0·042. These probabilities are not very small so it is wrong to come to any conclusions on correlation from one such small sample.

Notice the probabilities are symmetrically distributed about $\Sigma d^2 = 10$. Thus from one half of one of the tables we can read off the probabilities of obtaining values of $\Sigma d^2 \geqslant$ or $\leqslant$ any value of Σd^2. Thus, if we are given the following table,

Table 20.1

$\Sigma d^2 \geqslant$	10	12	14	16	18	20
P	0·542	0·458	0·375	0·208	0·167	0·042

the probability of obtaining a value of $\Sigma d^2 \geqslant 18$ from random sampling when $n = 4$ is 0·167 and the probability of obtaining a value of $\Sigma d^2 \leqslant 2$, that is, $20-18$, is also 0·167.

EXAMPLE

The corresponding rankings of two sample variables are:

First variable   1   2   3   4
Second variable 2   1   4   3

Calculate the rank correlation coefficient. Is there evidence of correlation between the variables?

SOLUTION

$$d = \quad -1 \quad 1 \quad -1 \quad 1$$
$$d^2 = \quad 1 \quad 1 \quad 1 \quad 1 \quad \text{and } \Sigma d^2 = 4$$

$$r_s = 1 - \frac{6 \times 4}{4(16-1)} = 0·6$$

If there is no correlation and $n = 4$, the expected value of Σd^2 is 10, and from Table 20.1 the probability of obtaining a value of $\Sigma d^2 \geqslant 16$ is 0·208. Therefore, the probability of obtaining a value of $\Sigma d^2 \leqslant (20-16)$, that is 4, is also 0·208, or about once in five. This is certainly not significant and there is no evidence of correlation between the two variables.

Larger Values of n

Now in a sample of size n, if $r_s = 0$, then $\Sigma d^2 = n(n^2-1)/6$, and if $r_s = 1$, then $\Sigma d^2 = 0$, and if $r_s = -1$, then $\Sigma d^2 = n(n^2-1)/3$, and in this case the probabilities are symmetrically distributed about $n(n^2-1)/6$ and the maximum and minimum values of d^2 are $n(n^2-1)/3$ and 0.

As n increases $n!$ increases very rapidly and for $n = 10$, $10! = 3\ 628\ 800$.

Sample range $n = 4$ to $n = 10$
The figures in Table 8 give the probability that $\Sigma d^2 \geqslant$ to a given value and cover the range $n = 4$ to $n = 10$ for selected values of Σd^2 ranging from about the 20% level to the 0·1% level of significance. Maximum values of Σd^2, that is $n(n^2-1)/3$ are given in the bottom row of the table and the mid-value of Σd^2, that is the value that makes $r_s = 0$, is half of this value.

EXAMPLE

Ten boys are ranked according to their ability in two subjects. Calculate Spearman's rank correlation coefficient. Is there evidence of correlation?

Boy	A	B	C	D	E	F	G	H	I	J
Rank in subject one	1	2	3	4	5	6	7	8	9	10
Rank in subject two	8	9	3	7	1	4	2	5	6	10

SOLUTION

$d =$	-7	-7	0	-3	4	2	5	3	3	0
$d^2 =$	49	49	0	9	16	4	25	9	9	0

and $\Sigma d^2 = 170$

$$r_s = 1 - \frac{6 \times 170}{10(100-1)} = -0.0303$$

Turning to Table 8 and the column headed $n = 10$ we see that if the rankings in the two subjects are independent, the expectation for Σd^2 is 165, half the figure in the bottom line of the table, while there is a probability of 0.235 that $\Sigma d^2 \geqslant 208$. The probability that $\Sigma d^2 \geqslant 170$ will therefore be greater than this and it is clear that r_s does not differ significantly from zero.

EXAMPLE

The rankings of a sample of two variables are as follows:

A	8	7	6	5	4	3	2	1
B	6	5	8	7	2	3	4	1

Is there evidence of correlation?

SOLUTION

d	2	2	-2	-2	2	0	-2	0
d^2	4	4	4	4	4	0	4	0

and $\Sigma d^2 = 24$

Turning to Table 8 and the column $n = 8$ we see the value of Σd^2 that makes $r_s = 0$ is 168/2, that is 84. Since our value of Σd^2 is less than this, we have the probability of getting a value of $\Sigma d^2 \leqslant 24$ is the same as the probability of getting a value of $\Sigma d^2 \geqslant (168-24) \geqslant 144$ that is 0.029, and this is significant at the 5% level and therefore there is some evidence of correlation.

$$r_s = 1 - \frac{6 \times 24}{8(64-1)} = 0.714$$

EXAMPLE

The rankings of corresponding variables in a bivariate distribution are given below. Is there evidence of correlation?

Variable A	1	2	3	4	5	6	7	8
Variable B	8	2	5	1	7	3	6	4

SOLUTION

$d =$	-7	0	-2	3	-2	3	1	4	
$d^2 =$	49	0	4	9	4	9	1	16	and $\Sigma d^2 = 92$

From the table the value of Σd^2 that makes $r_s = 0$ is 84, and the probability of $\Sigma d^2 \geqslant 108$ is 0·250. The value of Σd^2 in the example is between 84 and 108 and is not significant. Therefore there is no evidence of correlation.

$$r_s = -0·095$$

Sample range $n > 10$ to $n = 20$

For values of $n > 10$ the calculation of the distribution of Σd^2 becomes laborious, but it has been shown that in this interval r_s behaves like t with $n-2$ degrees of freedom, zero mean, and standard error

$$\sqrt{\left(\frac{1-r_s^2}{n-2}\right)}$$

i.e.,

$$t = \frac{r_s - 0}{\sqrt{\left(\frac{1-r_s^2}{n-2}\right)}} = r_s \sqrt{\left(\frac{n-2}{1-r_s^2}\right)}$$

EXAMPLE

Evaluate r_s for the following rankings. Is there evidence of correlation?

X	1	2	3	4	5	6	7	8	9	10	11	12
Y	12	11	8	10	3	9	7	6	2	1	5	4

SOLUTION

$d =$	-11	-9	-5	-6	2	-3	0	2	7	9	6	8
$d^2 =$	121	81	25	36	4	9	0	4	49	81	36	64

and $\Sigma d^2 = 510$

$$r_s = 1 - \frac{6 \times 510}{12 \times 143} = -0·783$$

$$\text{Standard error} = \sqrt{\left[\frac{1-(-0·783)^2}{10}\right]} = 0·197$$

Therefore,

$$t = \frac{-0·783}{0·197} = -3·97$$

with 10 degrees of freedom. This value is highly significant and indicates that negative correlation exists.

Sample range $n > 20$
For values of $n > 20$ it has been shown that an adequate test of significance may be obtained by assuming r_s is normally distributed with zero mean and standard error $1/\sqrt{(n-1)}$. Since Σd^2 is discrete, successive values differing by two units, there should be a continuity correction of plus or minus one, but since the numbers involved are large this correction can be omitted.

EXAMPLE

In a sample of size 101, $r_s = 0.3$. Is this value significant?

SOLUTION

Standard error $= 1/\sqrt{(n-1)} = 1/10$.
Normal variate $= (0.3 - 0)/0.1 = 3$.
From the normal variate table the probability of obtaining a value of the normal variate equal to or greater than this is $1 - 0.99865 = 0.00135$. This is highly significant and gives evidence of correlation.

Exercise 20.1

1   Evaluate r_s in each of the following cases. Find the probability of the answer being obtained by a random distribution of ranks, and state whether there is a significant indication of correlation.

(i) X   1 2 3 4 5 6
    Y   3 2 1 4 5 6

(ii) X   1 2 3 4 5 6 7 8
     Y   8 7 1 6 5 2 4 3

(iii) X   1 2 3 4 5 6 7 8 9 10
      Y   2 1 4 3 6 5 8 7 10 9

(iv) X   1 2 3 4 5 6 7
     Y   7 6 5 4 3 2 1

(v) X   1 2 3 4
    Y   2 1 4 3

2   Is there evidence of correlation in the following cases?

	(a)	(b)	(c)	(d)	(e)
r_s	0·5	0·7	0·4	0·3	−0·2
n	16	12	18	14	20

3   Is there evidence of correlation in the following bivariate distributions?

	(a)	(b)	(c)	(d)	(e)
r_s	0·3	0·5	0·1	0·4	0·2
n	50	101	82	65	26

4   The following fictitious data give the scores in an intelligence test and the reading rates of eight students:

Intelligence test	297	216	133	226	117	232	172	207
Reading rate	42	46	27	24	23	40	25	29

Find

(i) the product moment correlation coefficient between the scores,
(ii) the correlation coefficient by ranks.

Examine the hypothesis that there is no real correlation between the two scores.   (AEB)

5   What does the correlation coefficient measure?
    The table below gives the marks of 10 candidates in two examinations as a deviation from 50 marks.

Candidate	A	B	C	D	E	F	G	H	I	J
Examination 1	+27	−15	+19	+9	+5	−8	+36	+5	+11	−14
Examination 2	+7	−10	+7	−10	+16	−38	+15	−17	−35	−10

Calculate the coefficient of rank correlation between the two sets of marks.   (AEB)
    Is there evidence of correlation?

6   Nine candidates are placed in order of merit in English and French. The order in English paired with the order in French is shown below, together with the intelligence quotients of the candidates.

English	1	2	3	4	5	6	7	8	9
French	3	5	6	1	9	2	4	8	7
I.Q.	140	142	130	138	120	126	110	110	115

Calculate the coefficient of rank correlation between

(a) the order in English and the order in French,
(b) the order in English and the intelligence quotients,
(c) the order in French and the intelligence quotients.

    Explain what is meant by saying that a given coefficient of rank correlation is significant at the 5% level.   (L)
    Calculate the significant probabilities.

7   (a) Explain the use of (i) the product-moment correlation coefficient, (ii) the coefficient of correlation by ranks.

Give two examples of the appropriate application of each.

(b) At the final judging of a 'Cow of the Year Show', two judges gave the descending orders of merit of ten cows as E A H J B I F C G D and E H J A F I C B D G. Find the rank correlation coefficient between these two orders. Discuss the significance of the result.   (AEB)

The χ^2 Distribution

Distribution of χ^2

In Chapter 13 on Goodness of Fit the statistic χ^2 was defined as

$$\sum \frac{(x-\mu)^2}{\sigma^2}$$

and it was said that it could be reduced to

$$\sum \frac{(O-E)^2}{E}$$

where O was an observed result and E the estimated, or expected result. The statistic was used to estimate the goodness of fit of a curve. In the case of two cells it is easy to show the expression

$$\sum \frac{(O-E)^2}{E}$$

is distributed like χ^2. For example, suppose there is a mixed school and the chance of a pupil being a boy is p. Then the chance of a pupil being a girl is $q(=1-p)$. Suppose a random sample of n pupils is selected and let there be n_1 boys and n_2 girls. Then

$$n_1 + n_2 = n$$

and the number of degrees of freedom is $2-1 = 1$.

The expected number of boys and girls is np and nq, respectively. Tabulating the information, we have

	Observed	Expected
Boys	n_1	np
Girls	n_2	nq
Total	n	n

Therefore

$$\sum \frac{(O-E)^2}{E} = \frac{(n_1-np)^2}{np} + \frac{(n_2-nq)^2}{nq}$$

$$= \frac{(n_1-np)^2}{np} + \frac{[n-n_1-n(1-p)]^2}{nq}$$

$$= \frac{(n_1-np)^2}{np} + \frac{(n_1-np)^2}{nq} \quad [\text{Since } (-n_1+np)^2 = (n_1-np)^2]$$

$$= \frac{(n_1-np)^2(p+q)}{npq}$$

$$= \frac{(n_1-np)^2}{npq}$$

$$= \frac{(O_1-\mu)^2}{\sigma^2}$$

since $n_1 = O_1$ the observed result, $np = \mu = E$, the expected result, and $npq = \sigma^2$, the variance.

Size of Cell Frequencies

In problems we treat the cell frequencies as normal instead of binomial, that is as continuous instead of discrete, and this means the total frequency must be large—at least 30—and no expected frequency must be small. In some cases five can be accepted, but it is better if there is a minimum cell frequency of ten. If any cell frequency falls below this number two or more cells can be combined to bring the frequency to this number.

Use of χ^2 Table

The distribution of χ^2 is not symmetrical and the shape of the distribution depends on the value of v, the number of degrees of freedom.

The equation is $\qquad y = y_0 \chi^{v-2} \exp(-\chi^2/2)$

where y_0 is a constant depending on v such that the total area under the curve is unity.

Most tables of χ^2 give the percentage probability of exceeding a given value of χ^2, while others give the probability of reaching a given value of χ^2. That is, the probability values in the second table are the values in the first table subtracted from unity. The values in the table in the appendix

Distribution of χ^2

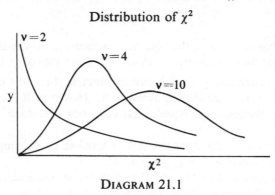

DIAGRAM 21.1

are the probability levels for exceeding a given value of χ^2. The probability of exceeding a given value of χ^2 is given by the shaded area in Diagram 21.2.

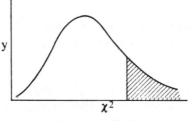

DIAGRAM 21.2

EXAMPLE

The graph of the χ^2 distribution with six degrees of freedom is shown in Diagram 21.3. Find the critical value for χ^2 for which

(i) the shaded area on the right is 0·05,
(ii) the total shaded area is 0·05,
(iii) the shaded area on the left is 0·05.

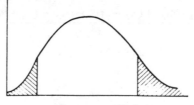

DIAGRAM 21.3

SOLUTION

(i)   In the table of χ^2 enter the column v and move downward to $v = 6$, then move right until under $P = 0.05$ or 5%. The value of χ^2 is then 12·59.

(ii) Since the distribution of χ^2 is not symmetric there are many critical values for which the total shaded area is 0·05. However, it is customary to make the two shaded areas equal; that is in our case each equals 0·025 or 2·5%.

The shaded area on the right is 0·025. Therefore proceeding as in (i) we have for $v = 6$ and $P = 2.5\%$, $\chi^2 = 14.45$.

The shaded area to the left is 0·025 and therefore the area to the right is 0·975 or 97·5%. Therefore for $v = 6$ and $P = 97.5\%$, $\chi^2 = 1.24$. Therefore the critical values are 1·24 and 14·45.

(iii) If the shaded area on the left is 0·05 the area to the right is 0·95. Therefore from the table for $v = 6$ and $P = 95\%$, $\chi^2 = 1.64$.

Exercise 21.1

1   From the table find the value of χ^2 for which the right-hand shaded area of Diagram 21.3 is 0·1 if the number of degrees of freedom is (i) 5, (ii) 10, (iii) 30, (iv) 70, (v) 100.

2   From the table find the value of χ^2 if the left-hand shaded area of Diagram 21.3 is 0·01 when the number of degrees of freedom is (i) 4, (ii) 12, (iii) 25, (iv) 50, (v) 95.

3   Calculate the critical values of χ^2 for which the total shaded areas are equal and v equals (i) 7, (ii) 23, (iii) 35, (iv) 60, (v) 75.

Confidence Limits for Variance

In a sample of size n drawn from a normal population of mean μ and standard deviation σ, let the sample mean and standard deviation be m and s, respectively. χ^2 for the sample will, by definition, be

$$\frac{(x_1-m)^2+(x_2-m)^2+ \ldots +(x_n-m)^2}{\sigma^2}$$

Now

$$\frac{\Sigma(x-m)^2}{n} = s^2$$

Therefore

$$\chi^2 = \frac{ns^2}{\sigma^2}$$

If $\chi^2_{0.025}$ and $\chi^2_{0.975}$ are the values of χ^2 for which 2·5% of the area lies in each tail of the distribution then the 95% confidence interval is

$$\chi^2_{0.025} > \frac{ns^2}{\sigma^2} > \chi^2_{0.975}$$

from which we estimate with 95% confidence that σ lies in the interval

$$\frac{s\sqrt{n}}{\chi_{0.975}} > \sigma > \frac{s\sqrt{n}}{\chi_{0.025}}$$

EXAMPLE

In a college of over 2000 male students a random sample of 49 students is taken. The mean height and standard deviation of the samples is 172·5 and 6·25 cm, respectively. Calculate 95% confidence limits for the standard deviation of all the students at the college.

SOLUTION

95% confidence limits are given by

$$\frac{s\sqrt{n}}{\chi_{0.975}} \quad \text{and} \quad \frac{s\sqrt{n}}{\chi_{0.025}}$$

for $(49-1)$ degrees of freedom.

By interpolation $\chi^2_{0.975}$ and $\chi^2_{0.025}$ are 30·8 and 69·0. Therefore, $\chi_{0.975} = \sqrt{30.8} = 5.55$ and $\chi_{0.025} = \sqrt{69.0} = 8.31$. The 95% confidence limits for the parent σ are therefore

$$\frac{6.25\sqrt{49}}{5.55} \quad \text{and} \quad \frac{6.25\sqrt{49}}{8.31}$$

or $\qquad$ 7·88 $\qquad$ and $\quad$ 5·26

EXAMPLE

A machine has been turning out 16 kg packages with a standard deviation of 0·1 kg. A random sample of 25 packages is taken and its standard deviation is calculated to be 0·14 kg. Is the increase in variability significant at the 0·05 level of significance?

SOLUTION

Assume the standard deviation of the parent population is 0·1 kg and make the hypothesis that there is no change in variability. The value of χ^2 for the sample is

$$\frac{ns^2}{\sigma^2} = \frac{25 \times (0\cdot14)^2}{(0\cdot1)} = 49$$

The value of $\chi_{0\cdot05}{}^2$ for a one-tailed test for $v = 24$ is 36·42. The calculated value of χ^2 is greater than this, and we therefore reject the hypothesis. There is evidence to show the variability has increased.

Exercise 21.2

1   A company manufactures electric light bulbs with a certain life. A random sample of 25 bulbs has a standard deviation of 140 hours. What are the 95% confidence limits for the standard deviation of all electric light bulbs manufactured?

2   A firm manufactures a rope and the standard deviation of its breaking strength is 533·8 N. It tries out a new process and the standard deviation of a sample of 10 ropes is 667·2 N. Is the increase in variability significant at the 5% level?

Confidence Limits for σ

The last examples have been testing the significant difference between s^2, the sample variance, and σ^2, the population variance. If n is small and we wish to test the difference between a sample estimate of the population variance, σ^2, and the population variance we must modify the formulae slightly.

$$\sigma_e{}^2 = \frac{ns^2}{n-1} \qquad\qquad (15.1)$$

or

$$s^2 = \frac{n-1}{n} \sigma_e{}^2$$

Therefore,

$$\chi^2 = \frac{\hbar s^2}{\sigma^2}$$

$$= \frac{(n-1)\sigma_e{}^2}{\sigma^2}$$

and the 95% confidence limits are

$$\frac{\sigma_e\sqrt{(n-1)}}{\chi_{0\cdot025}} < \sigma < \frac{\sigma_e\sqrt{(n-1)}}{\chi_{0\cdot975}}$$

Connection between χ^2 and F

$$\chi^2 = \frac{(n-1)\sigma_e^2}{\sigma^2}$$

or

$$\frac{\sigma_e^2}{\sigma^2} = \frac{\chi^2}{n-1}$$

which is a special case of F when $v_1 = n-1$ and $v_2 = \infty$. Thus the second worked example of the previous paragraph could be solved by the F-distribution.

Thus, $\sigma^2 = 0\cdot01$ and $s^2 = (0\cdot14)^2 = 0\cdot0196$.

Therefore $s^2/\sigma^2 = 1\cdot96$ where s has 24 degrees of freedom and σ an infinite number of degrees of freedom.

From the table for 5% values of F we get $F = 1\cdot52$ which is less than the calculated value. Therefore there is evidence of an increase in variability.

If we have two sample estimates of the variance

$$\sigma_{e_1}^2 = \frac{\chi_1^2}{n_1-1}\sigma^2$$

and

$$\sigma_{e_2}^2 = \frac{\chi_2^2}{n_2-1}\sigma^2$$

then

$$F = \frac{\sigma_{e_1}^2}{\sigma_{e_2}^2} = \frac{\chi_1^2}{\chi_2^2}\left(\frac{n_2-1}{n_1-1}\right)$$

EXAMPLE

In an examination it is found from past experience that the proportions of students obtaining 1st, 2nd, 3rd, and 4th grades are $1 : 3 : 4 : 5$. In the last examination 600 students were examined and graded as follows:

Grade	1st	2nd	3rd	4th
Frequency	50	120	179	251

Is there evidence of a change in standards?

SOLUTION

On the hypothesis that there is no change in standards the expected

frequencies are obtained by dividing the 600 students in the ratios
1 : 3 : 4 : 5. We then obtain the following table:

Grade	Observed frequency O	Expected frequency E	$O-E$	$(O-E)^2$	$\dfrac{(O-E)^2}{E}$
1st	50	46	4	16	0·348
2nd	120	138	−18	324	2·348
3rd	179	185	−6	36	0·195
4th	251	231	−20	400	1·732
					4·623

$$v = 4-1 = 3$$

$\chi^2 = 4\cdot623$ falls between the 10% and 95% levels of probability and does
not indicate a change in standards.

EXAMPLE

A coin was tossed 200 times and heads appeared 120 times. Is the coin
fair at the 1% level of significance?

SOLUTION

If the coin is fair the number of heads expected is 100. Arranging the
information in the form of a table, we have:

	O	E	$O-E$	$(O-E)^2$	$\dfrac{(O-E)^2}{E}$
Heads	120	100	20	400	4
Tails	80	100	−20	400	4
					8

$$v = 2-1 = 1$$

Entering the χ^2 table at $v = 1$ we get the 1% probability value as 6·63.
Therefore there is evidence that the coin is biased at the 1% level.

Yates' Correction for Continuity

The χ^2 distribution is a continuous distribution, but, as in the last example,
we often have to use the table for discrete values of $(O-E)^2/E$. We met a
similar difficulty when using the continuous normal distribution for discrete values of the variable. Thus when using the normal distribution to
solve problems involving the binomial distribution we had to make a

correction and 120 or more heads, say, was equivalent to 119·5 or more heads. In the χ^2 distribution if the frequencies are large the effect is not serious. Where the frequencies are not large, say between 5 and 10, a correction is applied in cases of one degree of freedom. The correction consists of subtracting 0·5 from each $(O-E)$ before squaring, and, in general, it is only found necessary to apply the correction in cases of one degree of freedom.

EXAMPLE

In the last example we were dealing with discrete variables and there was only one degree of freedom, but the frequencies were large, so the application of Yates' correction should not be really necessary as the following working illustrates.

			$O-E-0.5$	$(O-E-0.5)^2$	$\dfrac{(O-E-0.5)^2}{E}$
Heads	120	100	19·5	380·25	3·8025
Tails	80	100	−20·5	420·25	4·2025
					8·005

which gives the same result as before.

ALTERNATIVE SOLUTION

The probability of a head appearing is $\frac{1}{2}$, and therefore the mean number of heads expected is $200 \times \frac{1}{2} = 100$, with a standard deviation of $\sqrt{(200 \times \frac{1}{2} \times \frac{1}{2})}$ = 7·07.

In standard units and using the correction for continuity of the normal distribution, 120 or more heads is equivalent to

$$\frac{119\cdot5 - 100}{7\cdot07} = 2\cdot758$$

At the 1% level of significance the normal variate is 2·58, and our value falls outside this. Therefore there is evidence of bias at the 1% level of significance.

EXAMPLE

A card is drawn at random from a well-shuffled pack of cards and the suit noted. The card is then replaced and the pack again well-shuffled. This is repeated sixteen times and the numbers of each suit drawn are given below. Is there evidence to suggest the pack contains equal numbers of cards of each suit?

	Spades	Hearts	Diamonds	Clubs
	8	3	3	2

SOLUTION

If there are equal numbers of each suit we should expect four each. These frequencies are less than five so the χ^2 distribution will be in error. We therefore combine the frequencies so the evidence against the hypothesis is strongest. This we can do by combining Spades with, say, Hearts, and Diamonds with Clubs. The number of categories then becomes two, and the number of degrees of freedom one. We apply Yates' correction.

Category	O	E	$(O-E-0.5)$	$(O-E-0.5)^2$	$\dfrac{(O-E-0.5)^2}{E}$
Spades and Hearts	11	8	2.5	6.25	0.781
Diamonds and Clubs	5	8	−3.5	12.25	1.531

$$\chi^2 = 2.312$$

The value of χ^2 for $P = 10\%$ is 2.71 and our value of χ^2 is less than this. Therefore the results could easily have happened by chance from a pack with equal numbers of cards of each suit.

EXAMPLE

In 720 tosses of a pair of dice, 12 appeared 34 times, 6 appeared 98 times, and 2 appeared 8 times. Is there evidence the dice are biased?

SOLUTION

Assume the dice are not biased. A pair of dice can fall in $6 \times 6 = 36$ ways and 12 can appear in one way, 6 in five ways, and 2 in one way. Therefore the probability of a 12, 6, and 2 appearing is 1/36, 5/36, and 1/36, respectively.

In 720 throws we would expect

$$12 \text{ to appear } 720 \times 1/36 = 20 \text{ times,}$$
$$6 \text{ to appear } 720 \times 5/36 = 100 \text{ times,}$$
$$2 \text{ to appear } 720 \times 1/36 = 20 \text{ times.}$$

Tabulating, we have

Category	O	E	$O-E$	$(O-E)^2$	$\dfrac{(O-E)^2}{E}$
12	34	20	14	196	8·8
6	98	100	−2	4	0·04
2	8	20	−12	144	7·2
					16·04

For $v = 2$, $\chi^2 = 13\cdot81$ at the $0\cdot1\%$ level of significance. Therefore there is very strong evidence to assume the dice are biased.

Contingency Tables

When the total frequency of a distribution can be subdivided and tabulated in rows and columns the table is called a contingency table.

EXAMPLE

One hundred persons with newly-caught colds are divided into two groups, A and B, of fifty persons each. A special medicine is given to group A and water coloured and flavoured to resemble the medicine to group B. The persons are treated identically in every other respect. Thus each person thinks he is getting the same medical treatment. In three days 40 and 32 persons from groups A and B, respectively, recover from their colds. Is there evidence to suggest the medicine helps to cure colds?

SOLUTION

The data is tabulated in the table below.

	Observed frequencies		
	Recover	Do not recover	Total
Group A	40	10	50
Group B	32	18	50
Total	72	28	100

If we make the assumption that the medicine has no effect the expected frequencies are as in the following table.

	Expected frequencies		
	Recover	Do not recover	Total
Group A	36	14	50
Group B	36	14	50
Total	72	28	100

This table is made up on the assumption that 72 persons would have recovered in any case, and therefore we expect 36 in each group if the medicine is not effective. Notice that once we decide on the frequency for one of the four cells the frequencies for the other three cells are fixed. Thus we have only one degree of freedom. Using Yates' correction the calculation of χ^2 is:

$$\frac{(40-36-0\cdot5)^2}{36} + \frac{(32-36-0\cdot5)^2}{36} + \frac{(10-14-0\cdot5)^2}{14} + \frac{(18-14-0\cdot5)^2}{14}$$
$$= 3\cdot224$$

For $v = 1$ the value of χ^2 lies between the 10% and 5% probability levels. Therefore we assume there is no evidence to suggest the medicine is effective in curing colds.

Contingency Tables and Degrees of Freedom

The last example had a 2×2 contingency table. If there are m rows and n columns, the table is known as an $m \times n$ (read as m by n) contingency table. Because the row and column totals are fixed the number of degrees of freedom is obviously $(m-1)(n-1)$.

EXAMPLE

Five machines are used to manufacture articles which are graded into four categories.

Observed values

Grade	Machine					Total
	1st	2nd	3rd	4th	5th	
1st	750	620	500	680	650	3200
2nd	630	480	350	570	570	2600
3rd	120	80	130	170	100	600
4th	60	40	50	60	90	300
Total	1560	1220	1030	1480	1410	6700

Is there evidence of a difference between the quality of the product from the five machines?

SOLUTION

Assume there is no difference in the quality of the product from each machine. To obtain the expected frequencies estimate the ratios between the various grades to be proportionate to the grade totals. Thus for the first machine

the expected number of 1st grade articles is $\dfrac{3200}{6700} \times 1560 = 745$,

the expected number of 2nd grade articles is $\dfrac{2600}{6700} \times 1560 = 605$,

the expected number of 3rd grade articles is $\dfrac{600}{6700} \times 1560 = 140$,

the expected number of 4th grade articles is $\dfrac{300}{6700} \times 1560 = 70$.

Similarly for the 2nd, 3rd, 4th and 5th machines.

Grade	Expected values Machine					Total
	1st	2nd	3rd	4th	5th	
1st	745	583	492	707	673	3200
2nd	605	473	400	574	548	2600
3rd	140	109	92	133	126	600
4th	70	55	46	66	63	300
Total	1560	1220	1030	1480	1410	6700

Then $\chi^2 = \dfrac{(750-745)^2}{745} + \dfrac{(630-605)^2}{605} + \dfrac{(120-140)^2}{140} + \dfrac{(60-70)^2}{70} + \cdots$

and so on for all the cells = 72·538.

$$v = (5-1)(4-1) = 12$$

For $v = 12$ the value of $\chi^2 = 32·91$ at the 0·1% level of probability. Therefore there is ample evidence to show that the quality of the product differs in the five machines.

Additive Property of χ^2

Sometimes it happens that we have a number of tables of similar data from repeated experiments, and that the value of χ^2 for each experiment does not give a conclusive value for the significant probability. It is then natural to ask if we can obtain a more conclusive result by combining the different data. One way would be to pool the results and form a single table, but it can be shown that pooling may produce fallacious results. A better method is to add the different values of χ^2 and of v and to treat the final results as if they came from a single set of cells.

EXAMPLE

To test a hypothesis an experiment was repeated four times with the following results:

Experiment	1	2	3	4
χ^2	7·90	15·01	17·32	12·11
v	4	8	11	6

What conclusions can be drawn?

SOLUTION

The hypothesis cannot be rejected at the 5% level of significance for either experiment separately.

If the values of χ^2 and v are each added we get $\Sigma\chi^2 = 52\cdot34$ and $\Sigma v = 29$. For this value of v the probability of obtaining a value of $52\cdot34$ for χ^2 is $0\cdot5\%$, and so the hypothesis can be rejected.

Exercise 21.3

1   The last digit of 200 numbers from a local telephone directory was noted and the results were as follows:

Digit	0	1	2	3	4	5	6	7	8	9
Frequency	25	16	25	35	14	14	17	20	19	15

Is there evidence to assume they are a random selection of digits?

2   Out of 1000 persons 420 were inoculated against a certain disease. The following table gives the number of persons attacked and not attacked by the disease. Is there evidence of the effectiveness of the inoculation?

	Attacked	Not attacked
Inoculated	70	350
Not inoculated	80	500

3   A person having an article for sale advertised it in three newspapers. He received 15, 23 and 16 replies. Is there evidence that the second newspaper is the best paper to advertise in?

4   A die was thrown 240 times and the observed frequencies of the scores were as follows:

Scores	1	2	3	4	5	6
Frequency	33	48	40	41	29	49

Is this result consistent with the hypothesis that the die is unbiased?

5   A man tested four types of machines for a long period. The number of faults that developed in each type was 15, 9, 14 and 12. Is there evidence of differences in their performances?

6   Five hundred boys were asked to state whether they preferred Rugby, Soccer or Cricket and also which subject they preferred, Mathematics, Science or Languages. The results were as follows

	Mathematics	Science	Languages
Rugby	27	92	32
Soccer	14	71	43
Cricket	35	130	56

Are the results significant?

7   Out of 20 000 persons, 7000 were inoculated against a disease and the number attacked is given in the table below. Determine whether the attack is independent of inoculation.

	Attacked	Not attacked
Inoculated	106	6894
Not inoculated	372	12628

8   If in the last example 700 out of 2000 persons were inoculated and the numbers attacked were as follows, determine the new value of χ^2.

	Attacked	Not attacked
Inoculated	11	689
Not inoculated	37	1263

9   In a village A of 400 inhabitants 35% of the inhabitants said they attend church regularly, while in another village B of 600 inhabitants 40% of the inhabitants said they attended regularly. Is there a significant difference in the numbers attending church?

10   The standard deviation of the lifetimes of a sample of 10 manufactured articles is 100 hours. What are the 95% confidence limits for the standard deviation of all articles manufactured?

11   In a repeated experiment the following results were obtained:

χ^2	3·7	9·1	14·0	21·4
ν	1	4	7	13

What conclusions can be made?

12   In a large school the number of punishments and reprimands given during a certain week is listed below. Does the number of punishments vary with the day of the week?

Day	Mon.	Tues	Wed.	Thurs.	Fri.
Number of punishments	12	11	12	14	16

13   In a town, over nine years, the number of suicides in a month gave the following distribution:

x	0	1	2	3
f	61	41	5	1

where f is the number of months in which there are x suicides. Calculate the frequencies of the Poisson distribution having the same mean and total frequency. Is there evidence that the suicides do not occur independently?
(AEB)

14   At a public library, during a given week, the following numbers of books were borrowed

Day	Mon.	Tues.	Wed.	Thurs.	Fri.	Sat.
Number issued	204	292	242	283	252	275

Is there reason to believe that more books are generally borrowed on one weekday than on another?
(AEB)

15   Pupils from three schools A, B and C sit for an examination in Pure Mathematics and Applied Mathematics. The results are as follows:

	School			
	A	B	C	Total
Passed in both subjects	9	17	5	31
Passed in Pure Mathematics only	18	17	1	36
Passed in Applied Mathematics only	5	9	10	24
Failed in both subjects	6	2	1	9
Total	38	45	17	100

Examine the hypothesis (i) that there is no essential difference between the schools in Pure Mathematics, (ii) that there is no association between passing in Pure Mathematics and passing in Applied Mathematics, taking all the candidates together.
(AEB)

16   A table of random digits gives, for the first 200, the ten digits with the following frequencies:

0	1	2	3	4	5	6	7	8	9
19	18	16	26	21	24	19	22	16	19

Test the hypothesis that the digits appear with equal probability. Successive sets of 25 digits contain the following numbers of eights: 2, 0, 5, 2, 0, 6, 2, 5, 2, 3, 1, 1, 2, 2, 3, 7, 3, 2, 3, 3, 4, 2, 1, 1, 2, 3, 4, 2, 2, 3, 3, 0, 2, 1, 5, 5, 4, 2, 2, 3.
Use the Poisson distribution to find the probabilities of 0, 1, 2, . . . eights in a random set of 25 and compare observed with expected frequencies.   (AEB)

17 In three firms, the numbers of employees in three categories are as follows:

Firm	A	B	C
Skilled manual	28	42	50
Unskilled	33	76	71
Non-manual	39	82	79

Apply a suitable test to find whether there is a significant difference between the proportions of the various kinds of workers and state your conclusions.
(AEB)

18 The following fictitious data refer to hair and eye colours for 200 persons:

Eye colour	Hair colour Dark	Medium	Fair	Total
Blue	3	42	30	75
Grey	2	18	5	25
Brown	45	40	15	100
Total	50	100	50	200

Use the χ^2 test to examine the hypothesis that there is no association between hair and eye colours.   (AEB)

19 Oral tests are conducted by three examiners separately. The numbers of candidates in the categories credit, pass, fail are as shown in the table.

	Examiner A	B	C	Total
Credit	10	5	13	28
Pass	31	38	28	97
Fail	29	20	26	75
Total	70	63	67	200

Use a χ^2 test to examine the hypothesis that the examiners do not differ in their standards of awards.
State the assumptions made.   (AEB)

CHAPTER 22

Analysis of Variance

Purpose of the Method

If articles are manufactured or experiments performed the final result may be influenced by many factors. Consider the following fictitious example. Each of three laboratory assistants was asked to perform five experiments on a chemical. The results obtained were:

Assistant A	17	5	6	8	9
Assistant B	27	18	7	19	8
Assistant C	15	26	15	19	11

The final results could be affected by

  (i) the skill of the laboratory assistant,
 (ii) variations in the quality of the chemical, especially if purchased from different manufacturers,
(iii) faults in the apparatus used.

There may be other factors at work which could cause differences in the results obtained, and it is the purpose of 'Analysis of Variance' to find which of the factors, or combination of factors, has an appreciable effect on the final result, and to estimate the contribution each factor makes to the overall variability of the result. It must be realized that if none of these factors operates there will still be slight differences in the final results, which are due to unknowable causes, such as experimental and sampling errors, and this inevitable variation we assume to be normally distributed. It is known as the residual variation.

Single-factor Analysis

Let us simplify the problem mentioned in the first paragraph by assuming that the only cause that produces an effect on variability of the results other than the residual variation is the skill of the laboratory assistants. To test if this is so we make the hypothesis that there is no difference in the skill of each assistant, and that all the results come from the same parent population and are normally distributed. Therefore if we can calculate two independent values of the variance of the items they should be equal within the limits of sampling error. The two variabilities calculated are the variability of the parent population from within each assistant's results, and from the mean of each assistant's results.

The variability between the means measures the variability between each assistant plus chance effects, and the variability in each assistant's case must be due to chance effects only.

Calculation and Method

We will now illustrate the calculation and method using the example of the first paragraph assuming that there is no possibility of differences in the chemical, only in the skill of each laboratory assistant. The student is advised to study this example carefully because it utilises most of the theory of sampling.

CALCULATION [An explanation of the method is given after the calculation]

First tabulate the results as shown.

Experiment	Laboratory assistant		
	A	B	C
1	17	27	15
2	5	18	26
3	6	7	15
4	8	19	19
5	9	8	11

The calculation is as follows:

(i) Σx

	A	B	C
	45	79	86

(ii) m = mean of each assistant = $\Sigma x/n$ =

A	B	C
9	15·8	17·2

(iii) σ_e^2 = estimate of variability in each

$$\text{column} = \frac{\Sigma(x-m)^2}{n-1}$$

A	B	C
·90	278·76	128·8
4	4	4

(iv) Mean of the means = $\dfrac{\Sigma m}{3} = \bar{m} = \dfrac{9+15\cdot8+17\cdot2}{3} = 14$

(v) Variability between means = $\dfrac{(9-14)^2+(15\cdot8-14)^2+(17\cdot2-14)^2}{2}$

$$= 19\cdot24$$

(vi) Estimate of population variance from variance of means

$$= 5 \times 19 \cdot 24 = 96 \cdot 2 \quad \text{for } v = 2$$

(vii) Estimate of population variance from within sample variance

$$= \frac{\Sigma(x_1 - m_1)^2 + \Sigma(x_2 - m_2)^2 + \Sigma(x_3 - m_3)^2}{(n-1) + (n-1) + (n-1)}$$

$$= \frac{90 + 278 \cdot 76 + 128 \cdot 8}{4 + 4 + 4} = 41 \cdot 47 \quad \text{for } v = 12$$

(viii) Test the two values of the population variance for a significant difference using the F-distribution. $F_{1 \frac{2}{2}} = \dfrac{96 \cdot 2}{41 \cdot 47} = 2 \cdot 32$

From the tables $F_{1 \frac{2}{2}}$ for $P = 5\%$ is $3 \cdot 89$. Our value is well below this, so our hypothesis has not been disproved. That is, there is no evidence to show any difference in the skill of the laboratory assistants.

METHOD

(i) Σx for each assistant is obtained by adding the results of each of his five experiments.

(ii) The mean for each assistant is obtained by dividing each of the sums in (i) by 5.

(iii) An estimate of the population variance is found for each set of experiments by calculating $\Sigma(x - m)^2 / (n - 1)$ for each set of results. Remember we are estimating the population variance from a small sample and therefore divide by $(n - 1)$. In the case of assistant A the calculation is

$$\frac{(17-9)^2 + (5-9)^2 + (6-9)^2 + (8-9)^2 + (9-9)^2}{5 - 1} = \frac{90}{4}$$

(iv) Since each sample mean is an unbiased estimate of the population mean the best estimate we have of the population mean is $\Sigma m / 3$.

(v) We have a small sample of sample means and from this sample we wish to estimate the variability of the distribution of sample means. Therefore we divide by $(n - 1)$ that is $(3 - 1)$.

(vi) If σ^2 is the population variance, then the variance of the means of samples of size n is σ^2 / n. In our example there are five items in each sample. Therefore $\sigma^2 / 5 = 19 \cdot 24$, or, $\sigma^2 = 5 \times 19 \cdot 24 = 96 \cdot 2$ for two degrees of freedom.

(vii) The best estimate of the population variance to be obtained from sample variances is

$$\frac{(n_1-1)s_1{}^2+(n_2-1)s_2{}^2+\dots}{(n_1-1)+(n_2-1)+\dots}$$

that is,

$$\frac{\Sigma(x_1-m_1)^2+\Sigma(x_2-m_2)^2+\dots}{(n_1-1)+(n_2-1)+\dots}$$

(viii) What we wish to know is if the variability between the samples (assistant plus chance effect) is greater than the variability within the samples (chance effect). Hence we use the F-tables for a single-tail test of significance. If we wish to apply the z-test we proceed as follows:

$$z = \tfrac{1}{2}\log_e \frac{96\cdot2}{41\cdot46} = 0\cdot4208$$

which gives the same result.

In this example each assistant performed the same number of experiments. If the number of experiments had differed from assistant to assistant the analysis would be a little more complicated.

EXAMPLE

Four of each group of three different breeds of pigs were fed on the same ration for a period of time and the increase in mass of each pig was noted at the end of the period. Is there any significant difference in the sample means?

	Type of pig		
	A	B	C
Increase in mass (kg)	6	9	10
	8	8	10
	5	8	12
	5	7	8

SOLUTION

	A	B	C
(i) Σx	24	32	40
(ii) $m = \Sigma x/n$	6	8	10
(iii) $\sigma_e{}^2$ for each column	6/3	2/3	8/3

(iv) Mean of means $\qquad\qquad 24/3 = 8$

(v) Variability between means $= \dfrac{(6-8)+(8-8)+(10-8)}{2} = 4$

(vi) σ_e^2 from means $\qquad = 4 \times 4 = 16 \qquad$ for $v = 2$

(vii) σ_e^2 from within samples $\quad = \dfrac{6+2+8}{3+3+3} = 1\cdot78 \quad$ for $v = 9$

(viii) $F_9^2 = 16/1\cdot78 = 8\cdot99$. F_9^2 for 5% is $4\cdot26$. Therefore the value of F is significant. There is a difference.

Having decided our data are not homogeneous, that is, the three samples do not come from the same parent population, we wish to know which one, or more, of the differences between sample means may be significant. We can apply the *t*-test to the difference between means. The following procedure is best. The best estimate of the standard error of any sample mean is that calculated from the 'within samples' variance, since from this the variance due to the difference between means has been eliminated. This variance is $1\cdot78$ with $v = 9$. Therefore the standard error of any sample mean is $\sqrt{(1\cdot78/4)}$ since there are four items in a sample. The standard error of the difference of two means equals

$$\sqrt{\left(\frac{1\cdot78}{4} + \frac{1\cdot78}{4}\right)} = 0\cdot94$$

For $v = 9$, $t_{0\cdot05} = 2\cdot26$. Therefore

$$\frac{\text{difference between means}}{\text{standard error}} = 2\cdot26$$

for significance at the 5% level. Therefore the difference between means must be, for significance at 5% level, at least

$$2\cdot26 \times \text{standard error} = 2\cdot26 \times 0\cdot94 = 2\cdot12$$

Therefore, if the difference between sample means is equal to, or greater than $2\cdot12$, then they differ significantly at the 5% level. In our example the differences between the sample means is $A - B = -2$, $A - C = -4$, and $B - C = -2$. Therefore the difference between A and C is significant at the 5% level.

Simplification of the Magnitude of the Items

So far our examples have dealt with numbers of small magnitude. If the numbers are of large magnitude, they can be simplified by adding, subtracting, multiplying or dividing each item by a constant amount. Adding or subtracting does not affect the variance; if the numbers are all multiplied by n, the variance is multiplied by n^2, and if they are divided by n, the variance is divided by n^2.

EXAMPLE

Three operators make three readings for hardness on a block of metal. Determine whether there is any difference in the operators.

		Operator	
Experiment	A	B	C
1	715	720	718
2	713	717	719
3	714	719	718

SOLUTION

The mean of the nine results is 717. Subtract 717 from each member, thus coding the members.

		Operator	
Experiment	A	B	C
1	−2	3	1
2	−4	0	2
3	−3	2	1

Now proceed as before.

(i) Σx	−9	5	4
(ii) m	−3	5/3	4/3
(iii) σ_e^2 for each column	2/2	21/9	3/9
(iv) Mean of means	0		

(v) Variability between means $\dfrac{1}{2}\left(9+\dfrac{25}{9}+\dfrac{16}{9}\right)=\dfrac{61}{9}$

(vi) σ_e^2 from means $\dfrac{61}{3}$ for $v = 2$

(vii) σ_e^2 from within samples $\dfrac{66}{9 \times 6} = \dfrac{11}{9}$ for $v = 6$

(viii) $F_6^2 = \dfrac{61 \times 9}{3 \times 11} = 16 \cdot 64$

therefore highly significant.

Exercise 22.1

1 Four mixes of an alloy are made and for each mix five determinations of the density are found. The results are:

Determination	Mix			
	A	B	C	D
1	4·3	4·2	4·5	4·8
2	4·2	4·2	4·3	4·6
3	4·4	4·4	4·7	4·6
4	4·1	4·4	4·6	4·3
5	4·5	4·3	4·4	4·2

Is there evidence of variability in the mean densities?

2 Three analysts were asked to find the density of a mineral. Their results are shown in the table. Assuming the density of the mineral is constant, is there a significant difference between the skills of the analysts?

Experiment	Analyst		
	A	B	C
1	9	5	7
2	8	5	6
3	7	7	6
4	6	4	6
5	5	4	5

3 Six leaves were taken from each of four trees and their lengths measured in millimetres. Can the leaves be regarded as coming from the same species of tree?

	Tree			
	A	B	C	D
Leaf 1	84	89	90	87
Leaf 2	87	84	92	80
Leaf 3	90	91	89	86
Leaf 4	84	85	86	81
Leaf 5	85	91	85	82
Leaf 6	86	88	92	82

Use the t-test to find the significant difference at the 5% level for the difference between the means.

4 It is suggested that the temperature affects the yield of an operation. The temperature was maintained at different constant levels for four days, and three yields were obtained for each day. Is there evidence of temperature affecting yield? If there is, find between which days there is a significant difference at the 5% level.

	Day			
	1	2	3	4
Yield 1	33	46	30	43
Yield 2	32	53	33	47
Yield 3	28	51	39	45

5 The following dimensions are the diameters of shanks turned out by four machines. The results are in units of thousandths of a mm. Is there a difference between the shanks due to some peculiarity of the machines on which they are made?

	Machine		
A	B	C	D
8214	8224	8215	8205
8223	8230	8214	8192
8217	8194	8226	8237
8202	8211	8193	8209
8195	8222	8207	8190

6 The diameter of a tube is measured at five different points in each of four batches of tube. Can the batch means be regarded as constant?

	Batch			
	A	B	C	D
Diameter (tenths mm)	230	234	228	227
	230	232	232	229
	230	233	233	228
	229	231	228	233
	231	230	234	233

Further Simplification

So far the numerical work of the examples has been kept simple in order to allow the method to be understood. It is possible to derive simpler methods to calculate the expressions needed for the variabilities. It has been pointed out before that $\Sigma(x - \bar{x})^2 = \Sigma x^2 - (\Sigma x/n)^2$. Before we attempt to derive the simpler expressions let us prove a few identities. To avoid double suffixes let the value of the variable in column 1 be denoted by a, in column 2 by b, and so on.

Table

Row \ Column	1	2	3 ... r
1	a_1	b_1	c_1 ...
2	a_2	b_2	c_2 ...
n	a_n	b_n	c_n ...

Let
  (i) the sum of the squares of all items

$$= a_1^2 + a_2^2 + \ldots + a_n^2 + b_1^2 + b_2^2 + \ldots + b_n^2 + \text{etc.} = S.$$

(ii) the totals of each column 1 to r equal T_1, T_2, ... T_r,

(iii) $T_1 + T_2 + \ldots T_r = T$, the grand total,

(iv) the total number of items $= r \times n = N$,

(v) the means of the columns $= m_1$, m_2, ... m_r,

where $m_1 = \dfrac{\Sigma a}{n}$ etc. $= \dfrac{T_1}{n}$ etc.,

(vi) $T_1{}^2 + T_2{}^2 + \ldots + T_r{}^2 = \displaystyle\sum_{1}^{r} T_i^2$,

(vii) the sum of the means $= \dfrac{T_1}{n} + \dfrac{T_2}{n} + \ldots + \dfrac{T_r}{n} = \dfrac{T}{n}$,

(viii) the mean of the means $= \dfrac{T}{nr} = \dfrac{T}{N} = \bar{m}$

The sum of the squares of the deviations of the items of each column from the column mean equals

$$\Sigma(a - m_1)^2 + \Sigma(b - m_2)^2 + \ldots$$

$$= \Sigma a^2 - \frac{(\Sigma a)^2}{n} + \Sigma b^2 - \frac{(\Sigma b)^2}{n} + \ldots$$

$$= \Sigma a^2 + \Sigma b^2 + \ldots - \frac{1}{n}\left[(\Sigma a)^2 + (\Sigma b)^2 + \ldots\right]$$

$$= S - \frac{1}{n}\Sigma T_i^2$$

The sum of the squares of the deviations of the means from the grand mean

$$= \sum_{1}^{r} (m_i - \bar{m})^2 = \sum_{1}^{r} m_i{}^2 - \frac{(\Sigma m_i)^2}{r} = \frac{\Sigma T_i^2}{n^2} - \frac{T^2}{n^2 r}$$

The General Problem

Let us now use these identities to apply the analysis of variance to the general problem. Using Table 22.1 we have,

(i) $\Sigma x_i = T_1$, T_2, ... T_r.

(ii) $m_i = \dfrac{\Sigma x_i}{n} = \dfrac{T_1}{n}, \dfrac{T_2}{n}, \ldots \dfrac{T_r}{n}$.

(iii) $\sigma_e^2 =$ column estimate of variance $= \dfrac{\Sigma(a-m_1)^2}{n-1}, \dfrac{\Sigma(b-m_2)^2}{n-1} \ldots$

(iv) $\bar{m} = \dfrac{\Sigma m_i}{r} = \dfrac{\Sigma T_i}{nr} = \dfrac{T}{N}$

(v) Variability between means $= \dfrac{\Sigma(m_i-\bar{m})^2}{r-1} = \dfrac{1}{r-1}\left[\dfrac{\Sigma T_i^2}{n^2} - \dfrac{T^2}{n^2 r}\right]$

(vi) Estimation of population variance of means

$$= n\frac{\Sigma(m_i-\bar{m})^2}{r-1} = \frac{1}{r-1}\left[\frac{\Sigma T_i^2}{n} - \frac{T^2}{N}\right]$$

(vii) Estimation of population variance from within samples

$$= \frac{\Sigma(a-m_1)^2 + \Sigma(b-m_2)^2 + \ldots}{r(n-1)}$$

$$= \frac{1}{r(n-1)}\left[S - \frac{1}{n}\sum_1^r T_i^2\right]$$

(viii) Thus the two variances are calculated and can be tested by means of the F-distribution.

EXAMPLE

Let us now return to our first example and apply the new method. Tabulate as follows:

	Laboratory Assistant					
	A		B		C	
	a	a^2	b	b^2	c	c^2
	17	289	27	729	15	225
	5	25	18	324	26	676
	6	36	7	49	15	225
	8	64	19	361	19	361
	9	81	8	64	11	121
Totals	45	495	79	1527	86	1608

Therefore,

$$T = \Sigma T_i = 45+79+86 = 210$$
$$T^2 = 44\,100$$
$$S = 495+1527+1608 = 3630$$
$$\Sigma T_i^2 = 2025+6241+7396 = 15\,662$$
$$n = 5; \quad r = 3; \quad N = 15$$

Tabulating the results, we have

Variation	Sum of squares	Degrees of freedom v	Variance $\left[\dfrac{\text{Sum of squares}}{v}\right]$
Between columns	$\dfrac{\Sigma T_i^2}{n} - \dfrac{T^2}{N}$	$r-1$	
	$= \dfrac{15662}{5} - \dfrac{44100}{15}$	2	96·2
Residual (within columns)	$S - \dfrac{\Sigma T_i^2}{n}$	$r(n-1)$	
	$= 3630 - \dfrac{15662}{5}$	12	41·47

Therefore,

$$F_{12}^{2} = \frac{96\cdot2}{41\cdot47} = 2\cdot32 \text{ as before.}$$

EXAMPLE

A sample of ten machine components was taken from each of five machines. The length of each component, in cm, is given in the table. Assuming the lengths are normally distributed with constant variance, investigate if there is any difference between machines.

		Machine		
a	b	c	d	e
2·105	2·105	2·101	2·104	2·108
2·108	2·105	2·104	2·104	2·104
2·104	2·103	2·105	2·104	2·103
2·105	2·104	2·106	2·105	2·105
2·107	2·102	2·104	2·104	2·101
2·105	2·106	2·103	2·103	2·104
2·105	2·103	2·103	2·108	2·103
2·106	2·104	2·104	2·102	2·106
2·105	2·105	2·106	2·105	2·105
2·103	2·109	2·102	2·106	2·104

SOLUTION

Simplify the figures by multiplying by 1000 and subtracting 2100 from each. Arrange in the form of a table leaving places for the squares of the coded numbers.

a	a²	b	b²	c	c²	d	d²	e	e²
5	25	5	25	1	1	4	16	8	64
8	64	5	25	4	16	4	16	4	16
4	16	3	9	5	25	4	16	3	9
5	25	4	16	6	36	5	25	5	25
7	49	2	4	4	16	4	16	1	1
5	25	6	36	3	9	3	9	4	16
5	25	3	9	3	9	8	64	3	9
6	36	4	16	4	16	2	4	6	36
5	25	5	25	6	36	5	25	5	25
3	9	9	81	2	4	6	36	4	16

Totals	53	299	46	246	38	168	45	227	43	217

$$T = \Sigma T_i = 225$$
$$T^2 = 50\ 625$$
$$S = 299 + 246 + 168 + 227 + 217 = 1157$$
$$\Sigma T_i^2 = 2809 + 2116 + 1444 + 2025 + 1849 = 10\ 243$$
$$n = 10$$
$$r = 5$$
$$N = 50$$

Tabulating

Variation	Sum of Squares	Degrees of Freedom	Variance
Between columns	$\dfrac{10243}{10} - \dfrac{50625}{50}$	$5 - 1$	
	$= 11 \cdot 8$	$= 4$	$2 \cdot 95$
Between rows	$1157 - \dfrac{10243}{10}$	$5(10 - 1)$	
	$= 132 \cdot 7$	$= 45$	$2 \cdot 95$

$$F_{45}^4 = \frac{2 \cdot 95}{2 \cdot 95} = 1. \text{ Therefore not significant.}$$

Exercise 22.2

1   Rework the examples in Exercise 22.1 using the method of computation explained above.

2   Six machines produce components. The following data give the lengths

of samples of ten components from each machine. Test to see if the machines are constant.

(a)	12·2	12·0	12·1	12·7	12·4	12·2	12·2	12·8	12·2	12·3
(b)	12·1	12·7	12·5	12·9	12·6	12·6	12·3	12·8	12·6	12·1
(c)	12·0	12·2	12·1	12·0	12·8	12·6	12·1	12·0	12·5	12·1
(d)	12·5	12·5	12·9	12·2	12·4	12·2	12·5	12·7	12·4	12·4
(e)	12·9	12·0	12·4	12·4	12·9	12·0	12·3	12·0	12·4	12·1
(f)	12·0	12·3	12·3	12·5	12·3	12·7	12·0	12·1	12·1	12·5

3   In an examination in which the marks awarded formed approximately a normal distribution, there were ten candidates from each of three schools. The marks obtained were as follows:

School A	51	24	45	30	13	64	15	9	21	91
School B	94	43	80	22	99	27	84	76	97	60
School C	65	49	0	6	67	2	55	18	49	32

Carry out an analysis of variance, and examine the hypothesis that there is no essential difference between the schools.   (AEB)

An Observation

If we take the results we have obtained so far and tabulate, we have:

Source of variation	Sum of squares	Degrees of freedom
Between columns	$\dfrac{\Sigma T_i^2}{n} - \dfrac{T^2}{N}$	$r-1$
Residual (within columns)	$S - \dfrac{\Sigma T_i^2}{n}$	$r(n-1)$
If we add the columns we have	$S - \dfrac{T^2}{N}$	$rn-1 = N-1$

There is a third way of estimating the population variance by using all the items. The sum of the squares of the deviations from the mean is $\Sigma(x-\bar{m})^2$ which equals $\Sigma x^2 - (\Sigma x)^2/N$, that is $S - T^2/N$. This is the result we obtained above by adding. (*Note.* The variances are *not* additive).

Two-factor Analyses

If we return to our first problem we could modify it by assuming each laboratory assistant performed one experiment on each of five batches of the chemical, and then ask if there is any variation between the batches of chemical as well as between the skill of each laboratory assistant.

Tabulating the results, we get:

Batch	Laboratory Assistant A	B	C
1	17	27	15
2	5	18	26
3	6	7	15
4	8	19	19
5	9	8	11

We make the hypothesis that there is no difference between the skill of the laboratory assistants or between the batches of chemical. If in a similar way to one-factor analyses we calculate

  (i) a variance from the mean of columns, that is between the laboratory assistants,
 (ii) a variance from the means of rows, that is a batch variance,
(iii) a residual variance,
    they should be equal.

  (i) and (ii) are easy to calculate but (iii) is a little more difficult, but as in the case of one-factor analysis the sum of the Sum of Squares of (i), (ii) and (iii) is equal to the sum of the squares of the variance calculated from all the items (iv). Similarly for the degrees of freedom.

So it is usual to calculate (iii) by subtracting the Sum of the Squares and Degrees of Freedom of (i) and (ii) from (iv). Easy methods of computation can be deduced as in the case of one-factor analysis.

Tabulating the results, we have:

Row	Column 1	2	3 ...r
1	a_1	b_1	$c_1 \ldots$
2	a_2	b_2	$c_2 \ldots$
3	a_3	b_3	$c_3 \ldots$
n	a_n	b_n	$c_n \ldots$

$$S = a_1^2 + a_2^2 + \ldots + a_n^2 + b_1^2 + b_2^2 + \ldots + b_n^2 + \ldots$$

$T_1, T_2, \ldots T_r$ = totals of the columns

$R_1, R_2, \ldots R_n$ = totals of the rows

$$T_1 + T_2 + \ldots + T_r = R_1 + R_2 + \ldots + R_n = T$$

The total number of items is $rn = N$

$$T_1{}^2 + T_2{}^2 + \ldots T_r{}^2 = \sum_1^r T_i^2$$

$$R_1{}^2 + R_2{}^2 + \ldots R_n{}^2 = \sum_1^n R_i^2$$

Tabulating:

Source	Sum of Squares	Degrees of freedom
(i) Between columns	$\dfrac{\Sigma T_i^2}{n} - \dfrac{T^2}{N}$	$r-1$
(ii) Between rows	$\dfrac{\Sigma R_i^2}{r} - \dfrac{T^2}{N}$	$n-1$
(iii) Residual		
(iv) Totals	$S^2 - \dfrac{T^2}{N}$	$rn-1$

Adding (i) and (ii) and subtracting from (iv), we have

	Sum of squares	Degrees of freedom
(iii) Residual	$S - \dfrac{\Sigma T_i^2}{n} - \dfrac{\Sigma R_i^2}{r} + \dfrac{T^2}{N}$	$(n-1)(r-1)$

The student may ask why bother with (iii) when (i) and (ii) can be tested against (iv). If (i) or (ii) is suspect then (iv) which contains (i) and (ii) is also suspect, but (iii) does not depend on (i) or (ii) and still offers an estimate of the population variance.

EXAMPLE

Four chemists tested the strength of five batches of a chemical. The table gives the number of ml of standard tester used. Is there any difference between chemists or between the strengths of the batches of chemical?

Batch	Chemist			
	A	B	C	D
1	49	44	45	46
2	40	42	41	43
3	40	44	49	40
4	48	40	45	43
5	41	41	44	42

SOLUTION

Code the results by subtracting 40 from each item and tabulate as follows:

		A		B		C		D			Sum of
Batch	a	a^2	b	b^2	c	c^2	d	d^2	R	squares
1	9	81	4	16	5	25	6	36	24	158
2	0	0	2	4	1	1	3	9	6	14
3	0	0	4	16	9	81	0	0	13	97
4	8	64	0	0	5	25	3	9	16	98
5	1	1	1	1	4	16	2	4	8	22
Totals	18	146	11	37	24	148	14	58	67	389

$$T = R = 18 + 11 + 24 + 14 = 67$$
$$\Sigma T_i^2 = 324 + 121 + 576 + 196 = 1217$$
$$\Sigma R_i^2 = 576 + 36 + 169 + 256 + 64 = 1101$$
$$T^2 = 4489$$
$$S = 389$$
$$n = 5$$
$$r = 4$$
$$N = 20$$

Tabulating:

Source	Sum of squares	Degrees of freedom	Variance
Between columns	$\dfrac{1217}{5} - \dfrac{4489}{20} = 18 \cdot 95$	3	6·32
Between rows	$\dfrac{1101}{4} - \dfrac{4489}{20} = 53 \cdot 33$	4	13·33
Residual	92·27	12	7·69
Total	$389 - \dfrac{4489}{20} = 164 \cdot 55$	19	8·66

The sum of squares of the residual variance was obtained as follows, 164·55 $(18 \cdot 95 + 53 \cdot 33) = 92 \cdot 27$, and the residual number of degrees of freedom as $19 - (3 + 4) = 12$, and therefore the residual variance equals

$$\frac{92 \cdot 27}{12} = 7 \cdot 69$$

The F-numbers are:

between chemists; $F^{12}_3 = \dfrac{7 \cdot 69}{6 \cdot 32} = 1 \cdot 22.$ Not significant.

between batches; $F_{12}^4 = \dfrac{13 \cdot 33}{7 \cdot 69} = 1 \cdot 73.$ Not significant.

Exercise 22.3

1   The following data give the external diameters of a component made from four machines by five operators. Is there any 'between machines' or 'between operators' effect?

		Machine		
Operator	A	B	C	D
1	642	639	636	640
2	638	636	641	638
3	643	636	640	636
4	637	643	642	641
5	645	638	641	640

2   A chocolate manufacturer suspects there is something wrong with the masses of his 100 g bars of chocolate. He does not know if it is the mixes or the moulding machines that are at fault. The masses of samples from the four mixes and the four machines were as follows:

	Machine (mass in g)			
Mix	A	B	C	D
1	100·06	100·04	100·06	100·07
2	100·08	100·05	100·09	100·09
3	100·06	100·08	100·02	100·08
4	100·06	100·03	100·04	100·06

What conclusions can be drawn?

3   A bus company fitted three different makes of tyres on each of four buses. Four drivers were allotted one bus each and no one else was allowed to drive these buses. The lives of the tyres are shown below. Test for differences between tyres and between drivers.

	Driver			
Tyre	A	B	C	D
1	2600	2000	1400	1600
2	1400	1000	2000	2400
3	1600	1400	1400	2000

4   Four hospitals test three treatments for a disease. The number of days the patients take to recover are given below. Is there a difference between hospitals or between treatments?

	Treatment		
Hospital	A	B	C
1	20	17	25
2	17	21	24
3	22	10	25
4	23	15	10

5 Four different treatments were used on four different plots and the yield per plot in kg is given below. Is there a significant difference between plots or between treatments?

| Plot | Treatment | | | |
	A	B	C	D
1	283	321	398	263
2	264	268	348	222
3	301	300	362	272
4	324	355	384	245

6 An agricultural experiment was carried out by planting sixteen similar plots with four varieties of potatoes, and giving them four different manurial treatments, so that each variety and each treatment are used together in one plot.

The yields of potatoes in kilogrammes per plot were as follows:

| Treatment | Variety of potato | | | |
	A	B	C	D
1	236	244	260	267
2	302	301	316	339
3	316	320	334	347
4	296	309	321	316

Use analysis of variance to test for differences between varieties and between treatments. (AEB)

Further Analysis

The method of analysis of variance can be extended to three or more factors and has widespread application in many fields. One of the most important developments is the use of Latin squares, from which an analysis of three or four factors may be made on a small set of data. There is extensive literature on the subject.

Answers to the Exercises

Exercise 1.1, page 2

1   -3.75 2 34   3 $a = -2, b = 3$ 4 1 and 2   5 10

Exercise 1.2, page 2

(a) 4   (b) 1   (c) -6 (d) -1 (e) 14

Exercise 1.3, page 4

(a) 64   (b) $8\frac{1}{6}$ (c) $2\frac{24}{85}$ (d) 14π (e) 18.5

Exercise 1.4, page 6

1 (a) $x_1^2 + x_2^2 + x_3^2 + x_4^2$ (b) $x_1^3 y_1^2 + x_2^3 y_2^2 + x_3^3 y_3^2$
   (c) $x_1 + x_2 + x_3 + x_4 + 12$ (d) $3(x_1 + x_2 + x_3) + 3$
   (e) $3(x_1 + x_2 + x_3 + x_4 + x_5) - 4(y_1 + y_2 + y_3 + y_4 + y_5)$
   (f) $x_1^2 + x_2^2 + x_3^2 - 2(x_1 + x_2 + x_3) + 3$
   (g) $9(x_1^2 + x_2^2 + x_3^2 + x_4^2 + x_5^2) + 12(x_1 y_1 + x_2 y_2 + x_3 y_3 + x_4 y_4 + x_5 y_5)$
     $+ 4(y_1^2 + y_2^2 + y_3^2 + y_4^2 + y_5^2)$

2 (a) $\sum_1^7 x_r^2$ (b) $\sum_1^{10}(2x_r + y_r)$ (c) $\sum_1^n f_r x_r$ (d) $3\sum_1^n f_r x_r^2$ (e) $\sum_1^n f_r x_r^2 y_r$

3 (a) 34 (b) 228 (c) 48 (d) 540 (e) 1410

6 (a) $3n + \sum_1^n x_r$ (b) $an + b\sum_1^n x_r$ (c) $(a+b)\sum_1^n x_r$ (d) $\dfrac{n(n+1)}{2}$

   (e) $\dfrac{a(1-a^n)}{1-a}$ (f) $\dfrac{n^2 + 5n}{2}$ (g) $\dfrac{n+7}{2}$

Exercise 2.1, page 14

1 (a) (i) £19·5–£24·5; £24·5–£29·5; etc. (ii) £22; £27; etc. (iii) £5
    (iv) 0·0433; 0·0933; 0·1800; 0·2333; 0·1733; 0·1533; 0·0933;
    0·03

(b) (i) On a continuous scale: 0–50·5; 50·5–100·5; 100·5–150·5; 150·5–200·5; 200·5–250·5; 250·5–400·5; 400·5–450·5; 450·5–600·5

(ii) 25·25 (25·5 acceptable); 75·5; 125·5; 175·5; 225·5; 275·5; 425·5; 525·5

(iii) 50; 50; 50; 50; 50; 150; 150

(iv) 0·1538; 0·2692; 0·2154; 0·1481; 0·0962; 0·0769; 0·0173; 0·0231

(c) (i) 29·45–33·95; 33·95–38·45; etc.; to 74·45–

(ii) 31·70; 36·20; 40·70; etc.; to 72·20–

(iii) 4·50

(iv) 0·0035; 0·1304; 0·1970; 0·2414; 0·1869; 0·1422; 0·0609; 0·0212; 0·0127; 0·0035; 0·0024

(d) (i) 0·25–0·35; 0·35–0·45; etc. to 1·15–1·55

(ii) 0·30; 0·40; etc. . . . 1·35

(iii) 0·1; . . . 0·1; 0·4

(iv) 0·0498; 0·1099; 0·1806; 0·2068; 0·1605; 0·1265; 0·0755; 0·0463; 0·0230; 0·0209

(e) (i) 39·95–54·95; 54·95–59·95; . . .; 74·95–84·95; 84·95–120·05

(ii) 47·45; 57·45; 62·45; . . .; 79·95; 102·5

(iii) 15; 5; 5; 5; 5; 10; 35·1

(iv) 0·0290; 0·1859; 0·2233; 0·1936; 0·1587; 0·1434; 0·0661

Exercise 3.1, page 29

1   Arithmetic Mean   Median   Mode
(a)   17·2   9   9
(b)   49·5   48·0   44·6
(c)   42·8   42·7   42·5

2 (i) 14·55 (ii) 12·32 to 16·44 (iii) 68·16% (iv) 79·74%

3 By graph, about 6%

4 Mean = £18·1; Median = £20; Mode = £10. Only average altered is the mean. It becomes £18·3.

5 A.M. 49·5. Median 46·3. Marks grouped nearer 50 than 41. Paper II more difficult.

6 $A_L = 12·32$, $A_U = 13·46$

7 Mean = 57·22, taking open-ended class as 80·99. Median is better.

8 x 40 50 60 70 80 90 100
  f 2   8 12 17   7   4   2
  67·6, 57·7, 75·1

9 (i) 94·0, 102·6 (ii) 102·4

10 698, 696, 716 h
11 $-0\cdot4$, $-0\cdot9$, $+0\cdot3$, $+0\cdot7$
12 105·1
13 2, 0, 3, 6, 2, 7, 10, 9, 5, 4. 45·33 g; 45·14, 45·46 g.
14 -19, -4, $+2$, $+18$
15 (a) 141·1   (b) 134·1   (c) 131·2
16 47·4 km/h, 5 km/h, 6·25%
17 69, 52, 65, 69; 4·5, 18·6, 22·4, 36·1, 42·3, 47·0, 53·4

Exercise 4.1, page 46

1		Range	Mean deviation	$\frac{1}{2}(Q_2 - Q_1)$	σ
	(a)	97	16·56	3·75	12·4
	(b)	99	16·4	14	18·1
	(c)	45	7·07	6	8·9

2 Median $= 227$; A.M. $= 235\cdot1$; Variance $= 1075$
3 A.M. $= 46\cdot22$; $\sigma = 17\cdot1$
4 Mode $= 146\cdot6$; A.M. $= 152\cdot3$; $\sigma = 23\cdot1$
5 $f = 3, 15, 17, 35, 30, 36, 37, 15, 22, 20$. A.M. $= 130\cdot11$. $\sigma = 49\cdot92$
6 A.M. $= 20\cdot00$; $\sigma = 0\cdot012$
7 $Q_1 = 58$; $Q_2 = $ Median $= 58\cdot9$; $Q_3 = 59\cdot9$; $\frac{1}{2}(Q_8 - Q_1) = 0\cdot95$
8 A.M. $= 58\cdot96$. Mean deviation $= 1\cdot06$. $\sigma = 1\cdot36$
9 New mean $= 56\cdot25$. New $\sigma = 25$. New marks $= 34\cdot4$; 53·6; 87·2
10 A.M. $= 100\cdot5$; $\sigma = 13\cdot4$; Mean deviation $= 10\cdot13$
11 A.M. $= 3\cdot870$; $\sigma = 1\cdot92$
12 7255, 39, 33
13 5·14, 0·234%
14 35/9, 6·8

Exercise 5.1, page 54

(a) $y = 1\cdot1x + 16\cdot9$; $x = 0\cdot37y + 11\cdot39$
(b) $y = 0\cdot975x + 21\cdot118$; $x = 0\cdot984y - 19\cdot172$
(c) $y = 0\cdot973x + 6\cdot490$; $x = 0\cdot991y - 4\cdot448$
(d) $y = 0\cdot89x + 8\cdot20$; $x = 1\cdot001y - 3\cdot064$

Exercise 5.2, page 57

1 (a) $y = 0\cdot5x + 10$; $x = 1\cdot643y - 5\cdot712$
   (b) $y = 0\cdot980x + 20\cdot79$; $x = 0\cdot984y - 20$
   (c) $y = 0\cdot973x + 6\cdot49$; $x = 0\cdot99y - 4\cdot448$
   (d) $y = 0\cdot889x + 8\cdot196$; $x = 1\cdot009y - 3\cdot064$

Exercise 5.3 page 62

1　(i) $y = -0.7969x + 5.0594$; $x = -0.7590y + 4.7621$
　　(ii) $y = 0.6637x + 1.088$; $x = 0.7806y + 1.4264$
2　$y = 0.339x + 8.866$; $x = 1.518y + 41.332$
3　$b = 0.00088d^{2.60}$
4　$y = 0.669x + 79.558$; $x = 0.460y + 65.36$
5　$y = 16.876 - 1.108x$
6　$y = 0.592x + 1.446$; $x = 0.948y + 1.210$; 5.6 and 5.0
7　$y = 1.001x + 0.27$; $x = 0.916y + 3.28$
8　$x = 0.9945y + 1.480$; $x = -0.4246z + 12.82$; $y = -0.3182z + 10.82$;
　　10.70; 9.23; 6.45; 6.05; 10.66; 7.50
9　$y = -1.5x + 11.0$
10　(i) 28.79, 37.45　(ii) 39.73　(iii) $y = 1.06x + 2.79$
11　$y = 2.74x - 321$; 8, 14
12　$y = -1.126x + 28.46$; 8.8 thousand
13　$x = 28.4t + 176.8$; $y = 4.48t + 34.06$; 376, 65.4; 406, 699

Exercise 6.1, page 76

1　$y = 3.89 - 1.97x^2$
2　$R = 81.95 + 1.47v^2$
3　$y = 0.67 = \dfrac{1948}{x}$
4　$y = 1.67 + \dfrac{3.59}{x}$
5　$R = 172 + 0.365v^3$
6　(i) $y = 1.19 + 0.10x^2$　(ii) $y = 0.594 - 2.254x^2$
7　(i) $y = 2.74 + 1.4x + 0.7143x^2$
8　$y = -20.6 + 158.2x + 1.59x^2$; 220, 1720 sec.
9　$y = -2.14 + 3.07(x + x^2)$
10　$y = 3.155 + 1.977x^2$
11　$y = 2.74x - 321$
12　4, 1.6　(i) 23.4　(ii) 3.9

Exercise 7.1, page 85

1　From Ex. 5.3　(1) (i) -0.78; (ii) 0.72　2 0.72　3 0.93　4 0.55
2　-0.81　3 0.94　4 0.667　5 0.89　6 0　7 -0.844, High production
　　goes with low prices.　8 0.531　9 -0.091　10 0.629　11 $\bar{x} = -3$;
　　$\bar{y} = -2$, $r = -0.98$　12 (i) 0.424　(ii) $y = 0.287x + 0.736$　(iii) 0.97
　　13 (i) 11.375　(ii) $y = 3.25x - 7.5$　(iii) 0.94

14 (i) -0.40 (ii) -0.36
15 $y - 18.0 = 1.6(x + 1.9)$, $x + 1.9 = 0.29(y - 18.0)$; 0.68

Exercise 8.1, page 94

1 0.00248 2 (i) 0.0044 (ii) 0.105 3 (i) 0.23 (ii) 0.40 4 1.7p
5 £320

Exercise 8.2, page 100

1 (a)$\frac{1}{8}$ (b) $1/12$ 2 (a) 41   (b) 44   (c) 52   (d) 6   (e) 25
   (f) A $= 18$, B $= 20$, C $= 31$. Total $= 69$
3 A 2.84p   B 0.04p   C 2.79p
4 (i) $1/343$ (ii) $120/343$ $n = 4$
5 (i) $3/44$ (ii) $3/11$ (iii) $29/44$
6 (a) 179 to 1 against   (b) 8 to 7 against
7 (a) $1/63$ (b) $74/315$
8 (a) $1/12$ (b) $5/648$ (c) $31/216$
9 $p + p^2 + 2p^3$
10 $5/18$. 7. 2.416. $2\frac{1}{3}$. 1.71.
11 (i) $\dfrac{m}{m+n}$ (ii) $\dfrac{m(m-1)}{(m+n)(m+n-1)}$ (iii) $\dfrac{n}{m+n-1}$ (iv) $\dfrac{m}{m+r}$
13 (i) 0.9 (ii) 0.111 (iii) 0.336
    (iv) Large 0.336, Standard 0.320, Small 0.339. Ans. Small   (v) No
14 Chance of scoring more than $11 = 31/216$. More than $10 = 40/216$.
15 $2/27$ 16 $\frac{1}{3}$
17 (a) $\frac{1}{8}$ (b) $27/1000$ (c) $12/125$ (d) $3/25$ (e) $657/1000$
18 (i) $1/20$ (ii) $2/5$
19 (a) 0.904, 0.000008 (b) 0.56
20 (a) $1/12$, $1/12$ (b) 0.24
21 (i) $3/20$ (ii) $1/6$ (iii) $59/60$
22 (a) (i) $5/18$; (ii) $11/36$ (b) 0.107

Exercise 9.1, page 110

1 1260; 36; $2/7$; $2/3$
2 120; 28; 53 040   3 $5/21$ 4 576; 241 920   5 (a) 3136   (b) 2296. $\frac{3}{8}$
6 $\dfrac{12!}{5!4!2!}$; $\dfrac{9!4!}{12!}$ 7 65 780, 12 650, $\dfrac{1}{358\ 800}$
8 120; $\frac{1}{2}$ 10 (i) 0.0029 (ii) 88%
   11 (i) a $3/11$, b $3/44$ (ii) a $5/36$ b $5/18$
12 (i) a $\dfrac{1}{5525}$ b $\dfrac{6}{5525}$ (ii) 95.6%
13 $3(17)^{r-1}/(20)^r$ 14 (i) 15 120   (ii) 0.556 (iii) 0.0143

Exercise 10.1, page 121

1 0·599, 0·316, 0·075, 0·0105, 0·001
2 0·102 3 0·828 4 4 782 969; 0·00343 5 6 6 14

Exercise 10.2, page 123

1 117, 618, 1310, 1388, 735, 156
2 $p = 0·49$; 13·6, 52, 74·9, 48, 11·5
3 $p = 0·2143$; 114, 280, 306, 195, 80, 22, 4, 0, 0, 0
4 0, 2, 9, 20, 27, 24, 13, 4, 1
5 8, 2 (i) 0·0038 (ii) 4·914
6 (i) 1/221 (ii) 40/243
7 (i) $\frac{1}{8}, \frac{3}{8}$ (ii) a 3/32 b 35/128
9 11/144 10 $(1-q^3)^4(1+4q^3+10q^6)$
11 35·2s; 0·468 12 (i) 22; (ii) 56·6% 13 1, 0·335
14 79% 15 (i) 4·14% (ii) 95·7%, 8·5%
16 $\frac{m(m-n)}{2n^2} a^{m/n-2}b^2$, $\frac{m(m-n)(m-2n)}{6n^3} a^{m/n-3}b^3$, 4·000499953
17 2, 5% (i) 0·8718 (ii) 3; 4·52%
18 0·337 19 3, 2·7, 4

Exercise 11.1, page 143

1 0·247 2 26, 53, 54, 37, 19, 8, 3, 1, 0, 0
3 $P(0) = 0·6703$ $P(x \geqslant 2) = 0·062$
4 Mean = 1·93. Variance = 1·87. 15, 28, 27, 17, 8, 3, 1, 0.
5 (i) 0·018 (ii) 0·238 (iii) 0·265 6 10
7 4·85, 24·26, 60·64, 101·07, 126·34, 126·34, 105·28, 75·20, 47·00, 26·11,
   13·06, 5·93, 2·47, 0·95, 0·34
8 0·010 9 (i) 0·090 (ii) 0·053
10 0·0627, 0·1737, 0·2408, 0·2226, 0·1543, 0·0856, 6 or more 0·0603
11 11
12 2·0, 10·5, 27·9, 49·3, 65·5, 69·5, 61·5, 46·7, 31·0, 18·3, 9·7, 4·7, 3·4
13 0·30, 0·37, 0·22, 0·087, 0·026 14 0·84
15 0·998, 0·983, 0·94, 0·86, 0·73, 0·57, 0·41, 0·27, 0·17, 0·09
16 Mean = 6·7. Variance = 6·6.
   0·999, 0·99, 0·96, 0·89, 0·78, 0·65, 0·50, 0·34, 0·23, 0·13, 0·075
17 0·1848, 0·168
18 Mean = 1·384. 25, 34, 25, 10, 5. (Chart).
19 By Chart. 0·090 20 (a) 0·30327 (b) 0·09020
21 (i) 1 (ii) 0·846 (iii) 0·013 22 0·5, 1·44% 23 8

24   1·45   (a) 6%   (b) 33%
25   *np, npq, m, m*; 16·7%, 59·4%
26   (i) 0·61   (ii) 0·30   (iii) 0·076; 0·287

Exercise 12.1, page 154

1   (i) $y = 398·94 \exp[-(x-12)^2/8]$, 0·1336, 241·96, 53·98
    (ii) $y = 79·79 \exp[-(x-3)^2/50]$, 10·80, 66·65, 10·80
    (iii) $y = 263·82 \exp[-(x-10·3)^2/10·58]$, 62·66, 262·84, 0·003
    (iv) $y = 234·313 \exp[-(x+4)^2/18]$, 96·33, 96·33, 6·65.

Exercise 12.2, page 158

1   (i) 0·1, 0·06, 242, 317·4, 54·0, 1954·5
    (ii) 10·80, 22·75, 66·64, 274·3, 10·80, 977·25
    (iii) 62·17, 67·84, 353·5, 813·3, 0, 1521
    (iv) 96·33, 161·75. 96·63, 1509. 6·64, 6·70.
2   (i) 1197·4, 368·2   (ii) 250·2, 1347·2   (iii) 1456

Exercise 12.3, page 162

1   (i) $y = 137·57 \exp[-(x-19·12)^2/16·82]$
    (ii) a. 7, 32, 102, 209, 269, 221, 114, 37, 9.   b. 6, 30, 101, 213, 274, 223, 110, 35, 7.
2   (i) $y = 70·70 \exp[-(x-50)^2/63·69]$
    (ii) a. 4, 34, 150, 312, 312, 150, 34, 4.   b. 3, 30, 148, 319, 319, 148, 30, 3.
3   (i) $y = 7·96 \exp[-(x-219·72)^2/296·8]$
    (ii) a. 4, 4, 8, 13, 19, 25, 30, 31, 30, 26, 20 ,13, 8, 5, 2, 2.   b. 2, 4, 9, 13, 19, 25, 30, 32, 30, 26, 20, 14, 8, 5, 2, 1.
4   (i) $y = 124·84 \exp[-(x-5·38)^2/0·405]$
    (ii) a. 6, 14, 28, 37, 33, 17, 6, 2.   b. 4, 14, 29, 37, 31, 17, 6, 1.

Exercise 12.4, page 166

1   (i) 0·3821   (ii) 0·1087   (iii) 0·00003
2   0·0013   3 0·096   4 0·4114

Exercise 12.5, page 169

Question	1	2	3	4
Mean	19·2	49·8	220	5·38
Standard deviation	2·75	5·3	12·2	0·45

Exercise 12.6, page 171

1 1090 hours 2 13·4% 3 4, 5, 9, 13, 16, 18, 17, 13, 10, 4, 3, 2
4 $\mu = 45·833$; $\sigma = 9·259$ 5 (i) 0·1357 (ii) 0·01786 (iii) 0·6223
6 3, 8, 20, 36, 46, 26, 12, 4, 1 7 1·5165 8 1900
9 $\mu = 54$; $\sigma = 22·6$ 10 259 11 3 12 (i) 46·4 (ii) 19·4 (iii) 346
   (iv) 76 (v) 154 (vi) 467 13 (i) 0·1364 (ii) 0·00897 (iii) 0·7746
14 2% 15 4·041 and 3·959
16 (i) 0·3129 (ii) 0·0369 (iii) 0·8911 17 $\mu = 15·75$; $\sigma = 0·0523$
18 2 19 0·1165 20 (i) 110·8 (ii) 116·6 (iii) 120·5 (iv) 137. 2%
21 1·004375 22 (i) 0·3774 (ii) 0·05375
23 0·0088 24 7·8 pence 25 $\mu = 1·6175$; $\sigma = 0·0625$; 0·0425
26 (i) 16 (ii) 16 (iii) 16 (iv) 4·94. 26 27 3·011, 0·025; 7·5%
28 (i) 4·75% (ii) 6·68% (iii) 25% 29 49·80, 19·46 (a) 10%
   (b) 21·5% 30 3·289, 0·0364; 66% 31 1513, 24·5 m; 60·7%
32 (i) 0·62%; (ii) 4·8%; (iii) 8·2%; (iv) 42%
33 10·29½, 4 min (a) 9% (b) 21·5% 34 1·62 m, 7·2 cm, 1·84 m
35 0·16, 0·55%

Exercise 13.1, page 194

1 (i) $\chi^2 = 0·8982$; $v = 4$ (ii) $\chi^2 = 1·74$; $v = 3$ (iii) $\chi^2 = 36·42$;
   $v = 4$ (iv) $\chi^2 = 0·72$; $v = 3$
2 (i) $\chi^2 = 0·72$; $v = 3$ (ii) $\chi^2 = 1·56$; $v = 3$
3 (i) $\chi^2 = 2·68$; $v = 6$ (ii) $\chi^2 = 5·38$; $v = 4$
4 Too good to be true.

Exercise 14.1, page 201

1 (a) 0·00379 (b) 0·9735 2 (a) 0·7597 (b) 2·86 3 (a) 23·26,
   0·0079 4 (a) 23·26, 0·008 (b) 23·26, 0·0067 5 0·00621
6 Prob. = 0·00043. He is justified. 7 $n \geqslant 41$ 8 (a) 0·2635 (b) 0·0570
9 $n = 39$ 10 Prob. 0·00097. Improb.

Exercise 14.2, page 209

1 (a) 0·6065 (b) 0·00114 (c) 0·331 2 0·18 3 0·026 4 (a) 12·31
   (b) 0·71 5 0·020 6 Prob. 0·144. No. 7 $\mu = 7·6$, $\sigma = 0·018$
8 7·13% 9 0·0245 10 0·4 11 $X = 3·695$ Highly sig.
12 $X = 0·2084$. Not sig. 13 0·24 14 $X = 1·7$. Not sig.
15 $\mu = 17·2$, $\sigma = 0·96$ 16 29·45 and 28·95 17 Sig. inaccurate
18 $\mu = 112·1$ 19 (a) 0·48 (b) 0·43 (c) 0·49 20 0·3795, 0·040
21 Prob. = 0·042. Not sig. at 5%. Evidence of increase at 5% level.

22   Weak evidence of bias. Strong evidence of bias.
23   Using normal dis. 51 days; using binomial dis. 52 days. $P = 0.21$. N.S.
24   $X = -1.78$. Highly sig., 6 more needed.   25 $X = 1.55$. Not sig.
     Number <71.28 and >49.72 26 (i) 14.20, 0.04   (ii) 14.12 to 14.28
     (iii) Yes   27 (i) 1.6%   (ii) 2.989 to 3.001   (iii) 4.6%
28   (a) (i) 1097   (ii) 844   (b) 63.5   (c) No   (d) 713   29 6.169,
     0.00135
30   σ^2/n; 175.3$\pm$1.32; 175.3$\pm$1.56   31 Rectangular, from $-\frac{1}{2}$ to $\frac{1}{2}$.
     0, 1/12. Var y = Var $z - 1/12$ 32 4.8 standard errors, V.H.S. 66.24$\pm$
     1.11, 69.74$\pm$0.905   33 18.53, 3.817. Highly sig.
34   (i) No   (iii) 9, 15, 23, 24, 17, 4, 8

Exercise 14.3, page 223

1   (a) 6; 0.117   (b) 2.5; 0.039   (c) 6π; 0.072π (d) 6.75π; 0.086π
2   0.35π; 0.000185π
3   110; 5.52

Exercise 15.1, page 229

1   21.5534$\pm$0.0046   2 (i) 300$\pm$5.93   (ii) 300$\pm$4.51   (iii) 300$\pm$1.93
3   (a) (i) 355   (ii) 200   (b) (i) 614   (ii) 346   4 $X = 2.991$. No.
5   $X = 3.513$. Very sig. 714 to 686.   6 X for dif. of means 2.769
7   1380 to 1421   8 Prob. 0.093. N.S.

Exercise 15.2, page 239

1   (a) Reject the old process if the mean breaking strength of samples of
    100 fibres exceeds 204.7 N
    (b) The probability of accepting the old process when the mean
    breaking strength is 210 N is 0.004

2 | μ | 190 | 195 | 200 | 205 | 210 | 215 |
    |---|---|---|---|---|---|---|
    | β | 1.00 | 1.00 | 0.94 | 0.45 | 0.04 | 0.00 |

3   Accept the old drug if the number of cures in samples of 63 lies between
    44.16 and 3.64

4 | p | 0.1 | 0.2 | 0.3 | 0.4 | 0.5 | 0.6 | 0.7 | 0.8 | 0.9 |
    |---|---|---|---|---|---|---|---|---|---|
    | β | 0 | 0 | 0.002 | 0.236 | 0.806 | 0.236 | 0.002 | 0 | 0 |

Exercise 16.1, page 245

1   (i) 2.13   (ii) 2.05   (iii) 2.03   (iv) 1.96
2   (i) 2.23   (ii) 2.08   (iii) 2.03   (iv) 1.99

Exercise 16.2, page 249

1   57.4 and 50.6   2 1.83 and 1.79   3 $t = -2.655$. Sig. at 5%.

4 $t = 2\cdot31$. Yes. 5 No 6 Yes 7 No 8 Yes at 5%
9 No difference 10 Not sig. at 5% 11 Not sig. 12 6·51 and 11·83
13 $-3\cdot2$ to 7·4 14 Sig. at 10% not at 5% 15 $t = 3\cdot496$. Reject at 1%.
Paired observations at the same time. 16 0·0735 17 (i) Not sig.
(ii) Not sig. 18 Very sig. 19 Sig. at 2% 20 No. Use estimate
based on all heights.

Exercise 17.1, page 256

1 (i) 3·07 (ii) 2·90 (iii) 2·96 2 $F^{16}_8 = 1\cdot91$. Yes. 3 $F^{13}_9 = 2\cdot031$.
No. 4 $F^9 = 1\cdot3$. No. 5 $F^{11}_{11} = 2\cdot1$. No. 6 $F^{24}_{36} = 2\cdot115$ (i) Yes
(ii) No 7 $= 4\cdot49$. Sig. at $2\frac{1}{2}\%$ level. Type B more variable than
type A. 8 Variance ratio $= 1\cdot21$ with 9, 11 d.f. N.S. t-statistic $=$
$3\cdot92$ with 20 d.f. V.H.S. 9 (i) Estimate of F^9_6 is 5·5, just sig. at $2\frac{1}{2}\%$
level. Yes. (ii) The t-test is invalid in view of (i). Doubtful.

Exercise 18.1, page 267

1 (1) (a) $\mu = 5\cdot408$ (b) (i) $\sigma_e = 2\cdot47$ (ii) $\sigma_e = 2\cdot53$ (c) Inner 7·6
and 3·2. Outer 8·9 and 1·9. Inner. 3·92 and 0·33. Outer 4·96 and 0.
(2) (a) $\mu = 0\cdot1075$ (b) (i) $\sigma_e = 4\cdot97$ (ii) $\sigma_e = 4\cdot91$ (c) Inner 5·0
and $-4\cdot7$. Outer 7·7 and $-7\cdot5$. Inner 7·3 and 0. Outer 9·3 and 0.
2 (i) $\sigma_e^2 = 0\cdot00052$ (ii) $\sigma_e^2 = 0\cdot00050$. 5·6 and 5·7.
3 71·93 cm, 44·05 cm^2 (i) $71\cdot93 \pm 5\cdot49$ cm (ii) $71\cdot9 \pm 6\cdot71$ cm.
4 Warning line between 4 and 5. Action line between 6 and 7 (i) 0·55
(ii) 0·14. 0·55.

Exercise 19.1, page 275

1 (i) Yes (ii) No 2 (i) No (ii) No 3 (i) Yes (ii) No
4 (i) $0\cdot307 < P < 0\cdot801$ (ii) $0\cdot191 < P < 0\cdot841$ 5 Yes
6 $\bar{r} = 0\cdot691$, s.e. $= 0\cdot1141$ 7 For $v = 6$, sig. prob. $\simeq 10\%$. No evidence
of correlation

Exercise 20.1, page 282

1	r_s	Pro.	Sig.
(i)	0·771	0·051	Not sig.
(ii)	$-0\cdot57$	0·076	Not sig.
(iii)	0·9394	0·0011	Highly sig.
(iv)	0·6	0·208	Not sig.

2 (a) Doubtful (b) Highly sig. (c) Not sig. (d) Not sig. (e) Not sig.
3 (a) Sig. (b) Sig. (c) Not sig. (d) Sig. (e) Not sig.
4 (i) 0·672 (ii) 0·62. $\Sigma d^2 = 32$. Prob. not sig.
5 $\rho = 0\cdot45$. No.

6 (a) 11/30. Prob. 0·168. Not sig.   (b) 43/48. Prob. <0.004. Highly
   sig.   (c) 113/240. Prob <0.125. Not sig.
7 0·87. Sig.

Exercise 21.1, page 288

1 (i) 9·24   (ii) 15·99   (iii) 40·26   (iv) 85·53   (v) 118·5
2 (i) 0·297   (ii) 3·57   (iii) 11·52   (iv) 29·71   (v) 65·95
3 (i) 0·989, 20·28   (ii) 9·26, 44·18   (iii) 17·25, 60·25   (iv) 35·53,
   91·95   (v) 47·25, 110·25

Exercise 21.2, page 290

1 $111.6 < \sigma < 198.8$ 2 Increase in variability not sig.

Exercise 21.3, page 298

	χ^2	ν	Significance
1	19·90	9	Sig. at 2·5%
2	2·842	1	Sig. between 5 and 10%
3	2·111	2	Not sig.
4	7·9	5	Not sig.
5	1·68	3	Not sig.
6	6·840	4	Not sig.
7	35·401	1	Highly sig.
8	3·156	1	Sig. between 5 and 10%
9	2·56	1	Not sig.
10	$72.5 < \sigma < 192.5$		
11	$\Sigma\chi^2 = 48.2$	$\Sigma\nu = 25$	Neither at 5%, combined at 0·5
12	1·23	4	Not sig.
13	3·81	3	Not sig.
14	19·7	5	Highly sig.
15	(i) 12·03	2	Highly sig.
	(ii) 3·026	1	Evidence of association at 5%
16	4·80	9	Not sig.
	3·8	7	Good agreement
17	2·074	4	Not sig.
19	6·655 with 4 d.f. Not sig.		

Exercise 22.1, page 307

1 $F_{16}^{3} = 2.222$. Not sig. at 5%   2 $F_{12}^{2} = 3.333$. Not sig. at 5%.
3 $F_{20}^{3} = 5.46$. Very sig. Sig. dif. between B and D, C and D.
4 $F_{8}^{3} = 21.5$. Very sig. Sig. dif. between 1 and 2, 1 and 4, 3 and 4 at 5%.
5 $F^{16}_{3} = 2.66$. Not sig.   6 $F^{16}_{3} = 1.036$. Not sig. Yes.

Exercise 22.2, page 313

1 (1) $F_{16}^{3} = 2 \cdot 222$ (2) $F_{12}^{2} = 3 \cdot 333$ (3) $F_{20}^{3} = 5 \cdot 45$ (4) $F_{8}^{3} = 21 \cdot 5$
(5) $F_{3}^{16} = 2 \cdot 66$ (6) $F_{3}^{16} = 1 \cdot 036$
2 $F_{54}^{5} = 1 \cdot 68$. Not sig. 3 $F_{27}^{2} = 4 \cdot 93$. Sig at $2\frac{1}{2}\%$.

Exercise 22.3, page 318

1 Between machines $F_{3}^{12} = 1 \cdot 19$. N.S. Between operators $F_{4}^{12} = 1 \cdot 31$. N.S.
2 Between machines $F_{9}^{3} = 1 \cdot 62$. N.S. Between mixes $F_{9}^{3} = 1 \cdot 87$. N.S.
3 Between drivers $F_{3}^{6} = 1 \cdot 59$. N.S. Between tyres $F_{2}^{6} = 3 \cdot 02$. N.S.
4 Between treatments $F_{6}^{2} = 1 \cdot 50$. N.S. Between hospitals $F_{3}^{6} = 2 \cdot 06$. N.S.
5 Between treatments $F_{9}^{3} = 31 \cdot 02$. Highly sig. Between plots $F_{9}^{3} = 5 \cdot 92$. Sig.

Table 1   The Normal Distribution Function

x	Φ(x)	x	Φ(x)	x	Φ(x)	x	Φ(x)	x	Φ(x)
0·00	0·5000 40	0·50	0·6915 35	1·00	0·8413 25	1·50	0·9332 13	2·00	0·97725 53
·01	·5040 40	·51	·6950 35	·01	·8438 23	·51	·9345 12	·01	·97778 53
·02	·5080 40	·52	·6985 34	·02	·8461 24	·52	·9357 13	·02	·97831 51
·03	·5120 40	·53	·7019 35	·03	·8485 23	·53	·9370 12	·03	·97882 50
·04	·5160 39	·54	·7054 34	·04	·8508 23	·54	·9382 12	·04	·97932 50
0·05	0·5199 40	0·55	0·7088 35	1·05	0·8531 23	1·55	0·9394 12	2·05	0·97982 48
·06	·5239 40	·56	·7123 34	·06	·8554 23	·56	·9406 12	·06	·98030 47
·07	·5279 40	·57	·7157 33	·07	·8577 22	·57	·9418 11	·07	·98077 47
·08	·5319 40	·58	·7190 34	·08	·8599 22	·58	·9429 12	·08	·98124 45
·09	·5359 39	·59	·7224 33	·09	·8621 22	·59	·9441 11	·09	·98169 45
0·10	0·5398 40	0·60	0·7257 34	1·10	0·8643 22	1·60	0·9452 11	2·10	0·98214 43
·11	·5438 40	·61	·7291 33	·11	·8665 21	·61	·9463 11	·11	·98257 43
·12	·5478 39	·62	·7324 33	·12	·8686 22	·62	·9474 10	·12	·98300 41
·13	·5517 40	·63	·7357 32	·13	·8708 21	·63	·9484 11	·13	·98341 41
·14	·5557 39	·64	·7389 33	·14	·8729 20	·64	·9495 10	·14	·98382 40
0·15	0·5596 40	0·65	0·7422 32	1·15	0·8749 21	1·65	0·9505 10	2·15	0·98422 39
·16	·5636 39	·66	·7454 32	·16	·8770 20	·66	·9515 10	·16	·98461 39
·17	·5675 39	·67	·7486 31	·17	·8790 20	·67	·9525 10	·17	·98500 37
·18	·5714 39	·68	·7517 32	·18	·8810 20	·68	·9535 10	·18	·98537 37
·19	·5753 40	·69	·7549 31	·19	·8830 19	·69	·9545 9	·19	·98574 36
0·20	0·5793 39	0·70	0·7580 31	1·20	0·8849 20	1·70	0·9554 10	2·20	0·98610 35
·21	·5832 39	·71	·7611 31	·21	·8869 19	·71	·9564 9	·21	·98645 34
·22	·5871 39	·72	·7642 31	·22	·8888 19	·72	·9573 9	·22	·98679 34
·23	·5910 38	·73	·7673 31	·23	·8907 18	·73	·9582 9	·23	·98713 32
·24	·5948 39	·74	·7704 30	·24	·8925 19	·74	·9591 8	·24	·98745 33

x	Φ(x)	d	x	Φ(x)	d	x	Φ(x)	d	x	Φ(x)	d	x	Φ(x)	d
0·25	0·5987	39	0·75	0·7734	30	1·25	0·8944	18	1·75	0·9599	9	2·25	0·98778	31
·26	·6026	38	·76	·7764	30	·26	·8962	18	·76	·9608	8	·26	·98809	31
·27	·6064	39	·77	·7794	29	·27	·8980	17	·77	·9616	9	·27	·98840	30
·28	·6103	38	·78	·7823	29	·28	·8997	18	·78	·9625	8	·28	·98870	29
·29	·6141	38	·79	·7852	29	·29	·9015	17	·79	·9633	8	·29	·98899	29
0·30	0·6179	38	0·80	0·7881	29	1·30	0·9032	17	1·80	0·9641	8	2·30	0·98928	28
·31	·6217	38	·81	·7910	29	·31	·9049	17	·81	·9649	7	·31	·98956	27
·32	·6255	38	·82	·7939	28	·32	·9066	16	·82	·9656	8	·32	·98983	27
·33	·6293	38	·83	·7967	28	·33	·9082	17	·83	·9664	7	·33	·99010	26
·34	·6331	37	·84	·7995	28	·34	·9099	16	·84	·9671	7	·34	·99036	25
0·35	0·6368	38	0·85	0·8023	28	1·35	0·9115	16	1·85	0·9678	8	2·35	0·99061	25
·36	·6406	37	·86	·8051	27	·36	·9131	16	·86	·9686	7	·36	·99086	25
·37	·6443	37	·87	·8078	28	·37	·9147	15	·87	·9693	6	·37	·99111	23
·38	·6480	37	·88	·8106	27	·38	·9162	15	·88	·9699	7	·38	·99134	24
·39	·6517	37	·89	·8133	26	·39	·9177	15	·89	·9706	7	·39	·99158	22
0·40	0·6554	37	0·90	0·8159	27	1·40	0·9192	15	1·90	0·9713	6	2·40	0·99180	22
·41	·6591	37	·91	·8186	26	·41	·9207	15	·91	·9719	7	·41	·99202	22
·42	·6628	36	·92	·8212	26	·42	·9222	14	·92	·9726	6	·42	·99224	21
·43	·6664	36	·93	·8238	26	·43	·9236	15	·93	·9732	6	·43	·99245	21
·44	·6700	36	·94	·8264	25	·44	·9251	14	·94	·9738	6	·44	·99266	20
0·45	0·6736	36	0·95	0·8289	26	1·45	0·9265	14	1·95	0·9744	6	2·45	0·99286	19
·46	·6772	36	·96	·8315	25	·46	·9279	13	·96	·9750	6	·46	·99305	19
·47	·6808	36	·97	·8340	25	·47	·9292	14	·97	·9756	5	·47	·99324	19
·48	·6844	35	·98	·8365	24	·48	·9306	13	·98	·9761	6	·48	·99343	18
·49	·6879	36	·99	·8389	24	·49	·9319	13	·99	·9767	5	·49	·99361	18
0·50	0·6915		1·00	0·8413		1·50	0·9332		2·00	0·9772		2·50	0·99379	

Table 1   (cont.)

x	$\Phi(x)$	x	$\Phi(x)$	x	$\Phi(x)$	x	$\Phi(x)$
2·50	0·99379 17	2·65	0·99598 11	2·80	0·99744 8	2·95	0·99841 5
·51	·99396 17	·66	·99609 12	·81	·99752 8	·96	·99846 5
·52	·99413 17	·67	·99621 11	·82	·99760 7	·97	·99851 5
·53	·99430 16	·68	·99632 11	·83	·99767 7	·98	·99856 5
·54	·99446 15	·69	·99643 10	·84	·99774 7	·99	·99861 4
2·55	0·99461 16	2·70	0·99653 11	2·85	0·99781 7	3·0	0·99865 38
·56	·99477 15	·71	·99664 10	·86	·99788 7	3·1	·99903 28
·57	·99492 14	·72	·99674 9	·87	·99795 6	3·2	·99931 21
·58	·99506 14	·73	·99683 10	·88	·99801 6	3·3	·99952 14
·59	·99520 14	·74	·99693 9	·89	·99807 6	3·4	·99966 11
2·60	0·99534 13	2·75	0·99702 9	2·90	0·99813 6	3·5	0·99977 7
·61	·99547 13	·76	·99711 9	·91	·99819 6	3·6	·99984 5
·62	·99560 13	·77	·99720 8	·92	·99825 6	3·7	·99989 4
·63	·99573 12	·78	·99728 8	·93	·99831 5	3·8	·99993 2
·64	·99585 13	·79	·99736 8	·94	·99836 5	3·9	·99995 2
2·65	0·99598	2·80	0·99744	2·95	0·99841	4·0	0·99997

The function tabulated is $\Phi(x) = \dfrac{1}{\sqrt{2\pi}} \displaystyle\int_{-\infty}^{x} e^{-\frac{1}{2}t^{2}} dt$. $\Phi(x)$ is the probability that a random variable, normally distributed with zero mean and unit variance, will be less than x.

The critical table below gives on the left the range of values of x for which $\Phi(x)$ takes the value on the right, correct to the last figure given; in critical cases, take the upper of the two values of $\Phi(x)$ indicated.

x	$\Phi(x)$	x	$\Phi(x)$	x	$\Phi(x)$	x	$\Phi(x)$
3·075	0·9990	3·263	0·9995	3·731	0·99990	3·916	0·99995
3·105	0·9991	3·320	0·9996	3·759	0·99991	3·976	0·99996
3·138	0·9992	3·389	0·9997	3·791	0·99992	4·055	0·99997
3·174	0·9993	3·480	0·9998	3·826	0·99993	4·173	0·99998
3·215	0·9994	3·615	0·9999	3·867	0·99994	4·417	0·99999
							1·00000

Table 2   *The Normal Frequency Function*

x	$\phi(x)$	x	$\phi(x)$
0·0	0·3989	2·0	0·0540
0·1	·3970	2·1	·0440
0·2	·3910	2·2	·0355
0·3	·3814	2·3	·0283
0·4	·3683	2·4	0·224
0·5	0·3521	2·5	0·0175
0·6	·3332	2·6	·0136
0·7	·3123	2·7	·0104
0·8	·2897	2·8	·0079
0·9	·2661	2·9	·0060
1·0	0·2420	3·0	0·0044
1·1	·2179	3·1	·0033
1·2	·1942	3·2	·0024
1·3	·1714	3·3	·0017
1·4	·1497	3·4	·0012
1·5	0·1295	3·5	0·0009
1·6	·1109	3·6	·0006
1·7	·0940	3·7	·0004
1·8	·0790	3·8	·0003
1·9	·0656	3·9	·0002
2·0	0·0540	4·0	0·0001

The tabulation gives the ordinate $\phi(x) = \dfrac{1}{\sqrt{2\pi}}\,e^{-\frac{1}{2}x^2}$ of the normal frequency curve.

Table 3   *Percentage Points of the t-Distribution*

P	25	10	5	2	1	0·2	0·1	$\dfrac{120}{v}$
v = 1	2·41	6·31	12·71	31·82	63·66	318·3	636·6	
2	1·60	2·92	4·30	6·96	9·92	22·33	31·60	
3	1·42	2·35	3·18	4·54	5·84	10·21	12·92	
4	1·34	2·13	2·78	3·75	4·60	7·17	8·61	
5	1·30	2·02	2·57	3·36	4·03	5·89	6·87	
6	1·27	1·94	2·45	3·14	3·71	5·21	5·96	
7	1·25	1·89	2·36	3·00	3·50	4·79	5·41	
8	1·24	1·86	2·31	2·90	3·36	4·50	5·04	
9	1·23	1·83	2·26	2·82	3·25	4·30	4·78	
10	1·22	1·81	2·23	2·76	3·17	4·14	4·59	12
12	1·21	1·78	2·18	2·68	3·05	3·93	4·32	10
15	1·20	1·75	2·13	2·60	2·95	3·73	4·07	8
20	1·18	1·72	2·09	2·53	2·85	3·55	3·85	6
24	1·18	1·71	2·06	2·49	2·80	3·47	3·75	5
30	1·17	1·70	2·04	2·46	2·75	3·39	3·65	4
40	1·17	1·68	2·02	2·42	2·70	3·31	3·55	3
60	1·16	1·67	2·00	2·39	2·66	3·23	3·46	2
120	1·16	1·66	1·98	2·36	2·62	3·16	3·37	1
∞	1·15	1·64	1·96	2·33	2·58	3·09	3·29	0

The function tabulated is t_P defined by the equation

$$\frac{P}{100} = \frac{1}{\sqrt{v\pi}} \frac{\Gamma(\frac{1}{2}v+\frac{1}{2})}{\Gamma(\frac{1}{2}v)} \int\limits_{|t|\geqslant t_P} \frac{dt}{(1+t^2/v)^{\frac{1}{2}(v+1)}}.$$

If t is the ratio of a random variable, normally distributed with zero mean, to an independent estimate of its standard deviation based on v degrees of freedom, $P/100$ is the probability that $|t| \geqslant t_P$.

Interpolation v-wise should be linear in $120/v$.

Other percentage points may be found approximately, except when v and P are both small, by using the fact that the variable

$$y = \pm\sinh^{-1}(\sqrt{3t^2/2v}),$$

where y has the same sign as t, is approximately normally distributed with zero mean and variance $3/(2v-1)$.

Table 4 Transformation of the Correlation Coefficient

r	z	r	z	r	z
0·00	0·000 20	0·40	0·424 24	0·80	1·099 28
·02	·020 20	·42	·448 24	·81	·127 30
·04	·040 20	·44	·472 25	·82	·157 31
·06	·060 20	·46	·497 26	·83	·188 33
·08	·080 20	·48	·523 26	·84	·221 35
0·10	0·100 21	0·50	0·549 27	0·85	1·256 37
·12	·121 20	·52	·576 28	·86	·293 40
·14	·141 20	·54	·604 29	·87	·333 43
·16	·161 21	·56	·633 29	·88	·376 46
·18	·182 21	·58	·662 31	·89	·422 50
0·20	0·203 21	0·60	0·693 32	0·90	1·472 56
·22	·224 21	·62	·725 33	·91	·528 61
·24	·245 21	·64	·758 35	·92	·589 69
·26	·266 22	·66	·793 36	·93	·658 80
·28	·288 22	·68	·829 38	·94	·738
0·30	0·310 22	0·70	0·867 41		see next columns
·32	·332 22	·72	·908 42	0·95	1·832
·34	·354 23	·74	·950 46	·96	1·946
·36	·377 23	·76	0·996 49	·97	2·092
·38	·400 24	·78	1·045 54	·98	·298
0·40	0·424	0·80	1·099	·99	2·647
				1·00	∞

r	z	r	z	r	z	r	z
0·940	1·738 9	0·945	1·783 9	0·950	1·832 10	0·955	1·886 11
·941	·747 9	·946	·792 10	·951	·842 11	·956	·897 12
·942	·756 8	·947	·802 10	·952	·853 10	·957	·909 12
·943	·764 10	·948	·812 10	·953	·863 11	·958	·921 12
·944	·774 9	·949	·822 10	·954	·874 12	·959	·933 13
						0·960	1·946

r	z	r	z	r	z	r	z
0·960	1·946 13	0·965	2·014 15	0·970	2·092 18	0·975	2·185 20
·961	·959 13	·966	·029 15	·971	·110 17	·976	·205 22
·962	·972 14	·967	·044 16	·972	·127 19	·977	·227 22
·963	1·986 14	·968	·060 16	·973	·146 19	·978	·249 24
·964	2·000 14	·969	·076 16	·974	·165 20	·979	·273 25
						0·980	2·298

r	z	r	z	r	z	r	z
0·980	2·298 25	0·985	2·443 34	0·990	2·647 53	0·995	2·994
·981	·323 28	·986	·477 38	·991	·700 59	·996	3·106
·982	·351 29	·987	·515 40	·992	·759 67	·997	·250
·983	·380 30	·988	·555 44	·993	·826 77	·998	·453
·984	·410 33	·989	·599 48	·994	·903	·999	3·800
						1·000	∞

The function tabulated is $z = \tanh^{-1} r = \tfrac{1}{2}\log_e \dfrac{1+r}{1-r} = 1\cdot1513\log_{10}\dfrac{1+r}{1-r}$.

If r is a partial correlation coefficient, after s variables have been eliminated, in a sample of size n from a multivariate normal population with the corresponding partial correlation coefficient ρ, then z is approximately normally distributed with mean $\tanh^{-1}\rho + \rho/2(n-s-1)$ and variance $1/(n-s-3)$. For $s = 0$ we have the ordinary correlation coefficient.

Table 5   *Percentage Points of the χ^2-Distribution*

P	99·5	99	97·5	95	10	5	2·5	1	0·5	0·1
$\nu=1$	$0{\cdot}0^4393$	$0{\cdot}0^3157$	$0{\cdot}0^3982$	0·00393	2·71	3·84	5·02	6·63	7·88	10·83
2	0·0100	0·0201	0·0506	0·103	4·61	5·99	7·38	9·21	10·60	13·81
3	0·0717	0·115	0·216	0·352	6·25	7·81	9·35	11·84	12·84	16·27
4	0·207	0·297	0·484	0·711	7·78	9·49	11·14	13·28	14·86	18·47
5	0·412	0·554	0·831	1·15	9·24	11·07	12·83	15·09	16·75	20·52
6	0·676	0·872	1·24	1·64	10·64	12·59	14·45	16·81	18·55	22·46
7	0·989	1·24	1·69	2·17	12·02	14·07	16·01	18·48	20·28	24·32
8	1·34	1·65	2·18	2·73	13·36	15·51	17·53	20·09	21·95	26·12
9	1·73	2·09	2·70	3·33	14·68	16·92	19·02	21·67	23·59	27·88
10	2·16	2·56	3·25	3·94	15·99	18·31	20·48	23·21	25·19	29·59
11	2·60	3·05	3·82	4·57	17·28	19·68	21·92	24·73	26·76	31·26
12	3·07	3·57	4·40	5·23	18·55	21·03	23·34	26·22	28·30	32·91
13	3·57	4·11	5·01	5·89	19·81	22·36	24·74	27·69	29·82	34·53
14	4·07	4·66	5·63	6·57	21·06	23·68	26·12	29·14	31·32	36·12
15	4·60	5·23	6·26	7·26	22·31	25·00	27·49	30·58	32·80	37·70
16	5·14	5·81	6·91	7·96	23·54	26·30	28·85	32·00	34·27	39·25
17	5·70	6·41	7·56	8·67	24·77	27·59	30·19	33·41	35·72	40·79
18	6·26	7·01	8·23	9·39	25·99	28·87	31·53	34·81	37·16	42·31
19	6·84	7·63	8·91	10·12	27·20	30·14	32·85	36·19	38·58	43·82
20	7·43	8·26	9·59	10·85	28·41	31·41	34·17	37·57	40·00	45·31
21	8·03	8·90	10·28	11·59	29·62	32·67	35·48	38·93	41·40	46·80
22	8·64	9·54	10·98	12·34	30·81	33·92	36·78	40·29	42·80	48·27
23	9·26	10·20	11·69	13·09	32·01	35·17	38·08	41·64	44·18	49·73
24	9·89	10·86	12·40	13·85	33·20	36·42	39·36	42·98	45·56	51·18

ν									
25	10·52	11·52	13·12	14·61	34·38	37·65	40·65	44·31	46·93	52·62
26	11·16	12·20	13·84	15·38	35·56	38·89	41·92	45·64	48·29	54·05
27	11·81	12·88	14·57	16·15	36·74	40·11	43·19	46·96	49·64	55·48
28	12·46	13·56	15·31	16·93	37·92	41·34	44·46	48·28	50·99	56·89
29	13·12	14·26	16·05	17·71	39·09	42·56	45·72	49·59	52·34	58·30
30	13·79	14·95	16·79	18·49	40·26	43·77	46·98	50·89	53·67	59·70
40	20·71	22·15	24·43	26·51	51·81	55·76	59·34	63·69	66·77	73·40
50	27·99	29·71	32·36	34·76	63·17	67·50	71·42	76·15	79·49	86·66
60	35·53	37·48	40·48	43·19	74·40	79·08	83·30	88·38	91·95	99·61
70	43·28	45·44	48·76	51·74	85·53	90·53	95·02	100·4	104·2	112·3
80	51·17	53·54	57·15	60·39	96·58	101·9	106·6	112·3	116·3	124·8
90	59·20	61·75	65·65	69·13	107·6	113·1	118·1	124·1	128·3	137·2
100	67·33	70·06	74·22	77·93	118·5	124·3	129·6	135·8	140·2	149·4

The function tabulated is χ_P^2 defined by the equation $\dfrac{P}{100} = \dfrac{1}{2^{\nu/2}\Gamma(\frac{1}{2}\nu)}\displaystyle\int_{\chi_P^2}^{\infty} x^{\frac{1}{2}\nu-1}e^{-x/2}dx$. If x is a variable distributed as χ^2 with ν degrees of freedom, $P/100$ is the probability that $x \geq \chi_P^2$. For $\nu < 100$, linear interpolation in ν is adequate. For $\nu > 100$, $\sqrt{2\chi^2}$ is approximately normally distributed with mean $\sqrt{2\nu-1}$ and unit variance, and the percentage points may be obtained from Table 2.

Table 6(a)   *5 Per cent Points of the F-Distribution*

$v_1 =$	1	2	3	4	5	6	7	8	10	12	24	∞
$v_2 = 1$	161·4	199·5	215·7	224·6	230·2	234·0	236·8	238·9	241·9	243·9	249·0	254·3
2	18·5	19·0	19·2	19·2	19·3	19·3	19·4	19·4	19·4	19·4	19·5	19·5
3	10·13	9·55	9·28	9·12	9·01	8·94	8·89	8·85	8·79	8·74	8·64	8·53
4	7·71	6·94	6·59	6·39	6·26	6·16	6·09	6·04	5·96	5·91	5·77	5·63
5	6·61	5·79	5·41	5·19	5·05	4·95	4·88	4·82	4·74	4·68	4·53	4·36
6	5·99	5·14	4·76	4·53	4·39	4·28	4·21	4·15	4·06	4·00	3·84	3·67
7	5·59	4·74	4·35	4·12	3·97	3·87	3·79	3·73	3·64	3·57	3·41	3·23
8	5·32	4·46	4·07	3·84	3·69	3·58	3·50	3·44	3·35	3·28	3·12	2·93
9	5·12	4·26	3·86	3·63	3·48	3·37	3·29	3·23	3·14	3·07	2·90	2·71
10	4·96	4·10	3·71	3·48	3·33	3·22	3·14	3·07	2·98	2·91	2·74	2·54
11	4·84	3·98	3·59	3·36	3·20	3·09	3·01	2·95	2·85	2·79	2·61	2·40
12	4·75	3·89	3·49	3·26	3·11	3·00	2·91	2·85	2·75	2·69	2·51	2·30
13	4·67	3·81	3·41	3·18	3·03	2·92	2·83	2·77	2·67	2·60	2·42	2·21
14	4·60	3·74	3·34	3·11	2·96	2·85	2·76	2·70	2·60	2·53	2·35	2·13
15	4·54	3·68	3·29	3·06	2·90	2·79	2·71	2·64	2·54	2·48	2·29	2·07
16	4·49	3·63	3·24	3·01	2·85	2·74	2·66	2·59	2·49	2·42	2·24	2·01
17	4·45	3·59	3·20	2·96	2·81	2·70	2·61	2·55	2·45	2·38	2·19	1·96
18	4·41	3·55	3·16	2·93	2·77	2·66	2·58	2·51	2·41	2·34	2·15	1·92
19	4·38	3·52	3·13	2·90	2·74	2·63	2·54	2·48	2·38	2·31	2·11	1·88
20	4·35	3·49	3·10	2·87	2·71	2·60	2·51	2·45	2·35	2·28	2·08	1·84
21	4·32	3·47	3·07	2·84	2·68	2·57	2·49	2·42	2·32	2·25	2·05	1·81
22	4·30	3·44	3·05	2·82	2·66	2·55	2·46	2·40	2·30	2·23	2·03	1·78
23	4·28	3·42	3·03	2·80	2·64	2·53	2·44	2·37	2·27	2·20	2·00	1·76
24	4·26	3·40	3·01	2·78	2·62	2·51	2·42	2·36	2·25	2·18	1·98	1·73
25	4·24	3·39	2·99	2·76	2·60	2·49	2·40	2·34	2·24	2·16	1·96	1·71
26	4·23	3·37	2·98	2·74	2·59	2·47	2·39	2·32	2·22	2·15	1·95	1·69
27	4·21	3·35	2·96	2·73	2·57	2·46	2·37	2·31	2·20	2·13	1·93	1·67
28	4·20	3·34	2·95	2·71	2·56	2·45	2·36	2·29	2·19	2·12	1·91	1·65
29	4·18	3·33	2·93	2·70	2·55	2·43	2·35	2·28	2·18	2·10	1·90	1·64

30	4·17	3·32	2·92	2·69	2·53	2·42	2·33	2·27	2·16	2·09	1·89	1·62
32	4·15	3·29	2·90	2·67	2·51	2·40	2·31	2·24	2·14	2·07	1·86	1·59
34	4·13	3·28	2·88	2·65	2·49	2·38	2·29	2·23	2·12	2·05	1·84	1·57
36	4·11	3·26	2·87	2·63	2·48	2·36	2·28	2·21	2·11	2·03	1·82	1·55
38	4·10	3·24	2·85	2·62	2·46	2·35	2·26	2·19	2·09	2·02	1·81	1·53
40	4·08	3·23	2·84	2·61	2·45	2·34	2·25	2·18	2·08	2·00	1·79	1·51
60	4·00	3·15	2·76	2·53	2·37	2·25	2·17	2·10	1·99	1·92	1·70	1·39
120	3·92	3·07	2·68	2·45	2·29	2·18	2·09	2·02	1·91	1·83	1·61	1·25
∞	3·84	3·00	2·60	2·37	2·21	2·10	2·01	1·94	1·83	1·75	1·52	1·00

The function tabulated in Table 6 is F_P defined by the equation

$$\frac{P}{100} = \frac{\Gamma(\frac{1}{2}v_1 + \frac{1}{2}v_2)}{\Gamma(\frac{1}{2}v_1)\Gamma(\frac{1}{2}v_2)} v_1^{\frac{1}{2}v_1} v_2^{\frac{1}{2}v_2} \int_{F_P}^{\infty} \frac{F^{\frac{1}{2}v_1 - 1}}{(v_2 + v_1 F)^{\frac{1}{2}(v_1 + v_2)}} \, dF,$$

with $P = 5, 2\frac{1}{2}, 1$ and 0.1. If F is the ratio of a mean square on v_1 degrees of freedom to an independent mean square on v_2 degrees of freedom, and if the mean squares have equal expectations, then $P/100$ is the probability that $F \geq F_P$. The lower percentage points, that is the value F_P' such that $P/100$ is the probability that $F \leq F_P'$, may be found by interchanging v_1 and v_2 and using the reciprocal of the tabulated value.

Linear interpolation will usually be sufficiently accurate except when either $v_1 > 12$ or $v_2 > 40$, though occasionally a slight improvement may be effected by using harmonic interpolation. Otherwise, except when v_1 and v_2 are both large, interpolation should be linear in $v_1 F_P$ or $v_2 F_P$ (this is equivalent to harmonic interpolation). When v_1 and v_2 are both large the percentage points may be found from the formula

$$1.1513 \log_{10} F_P = \tfrac{1}{2} \log_e F_P = \frac{x_P \sqrt{h + \lambda}}{h} - \left(\frac{1}{v_1 - 1} - \frac{1}{v_2 - 1} \right)\left(\lambda + \frac{5}{6}\right).$$

where x_P is the P-per cent point of the normal distribution (Table 2), $\lambda = \frac{1}{6}(x_P^2 - 3)$ and $\frac{2}{h} = \frac{1}{v_1 - 1} + \frac{1}{v_2 - 1}$.

For the values of P given in Table 7, x_P and λ are as follows:

P	5	$2\frac{1}{2}$	1	0.1
x_P	+1·6449	1·9600	2·3263	3·0902
λ	−0·0491	+0·1402	0·4020	1·0916

Table 6(b)   1 Per Cent Points of the F-Distribution

$\nu_1 =$	1	2	3	4	5	6	7	8	10	12	24	∞
$\nu_2 = 1$	4052	5000	5403	5625	5764	5859	5928	5981	6056	6106	6235	6366
2	98·5	99·0	99·2	99·2	99·3	99·3	99·4	99·4	99·4	99·4	99·5	99·5
3	34·1	30·8	29·5	28·7	28·2	27·9	27·7	27·5	27·2	27·1	26·6	26·1
4	21·2	18·0	16·7	16·0	15·5	15·2	15·0	14·8	14·5	14·4	13·9	13·5
5	16·26	13·27	12·06	11·39	10·97	10·67	10·46	10·29	10·05	9·89	9·47	9·02
6	13·74	10·92	9·78	9·15	8·75	8·47	8·26	8·10	7·87	7·72	7·31	6·88
7	12·25	9·55	8·45	7·85	7·46	7·19	6·99	6·84	6·62	6·47	6·07	5·65
8	11·26	8·65	7·59	7·01	6·63	6·37	6·18	6·03	5·81	5·67	5·28	4·86
9	10·56	8·02	6·99	6·42	6·06	5·80	5·61	5·47	5·26	5·11	4·73	4·31
10	10·04	7·56	6·55	5·99	5·64	5·39	5·20	5·06	4·85	4·71	4·33	3·91
11	9·65	7·21	6·22	5·67	5·32	5·07	4·89	4·74	4·54	4·40	4·02	3·60
12	9·33	6·93	5·95	5·41	5·06	4·82	4·64	4·50	4·30	4·16	3·78	3·36
13	9·07	6·70	5·74	5·21	4·86	4·62	4·44	4·30	4·10	3·96	3·59	3·17
14	8·86	6·51	5·56	5·04	4·70	4·46	4·28	4·14	3·94	3·80	3·43	3·00
15	8·68	6·36	5·42	4·89	4·56	4·32	4·14	4·00	3·80	3·67	3·29	2·87
16	8·53	6·23	5·29	4·77	4·44	4·20	4·03	3·89	3·69	3·55	3·18	2·75
17	8·40	6·11	5·18	4·67	4·34	4·10	3·93	3·79	3·59	3·46	3·08	2·65
18	8·29	6·01	5·09	4·58	4·25	4·01	3·84	3·71	3·51	3·37	3·00	2·57
19	8·18	5·93	5·01	4·50	4·17	3·94	3·77	3·63	3·43	3·30	2·92	2·49

20	8·10	5·85	4·94	4·43	4·10	3·87	3·70	3·56	3·37	3·23	2·86	2·42
21	8·02	5·78	4·87	4·37	4·04	3·81	3·64	3·51	3·31	3·17	2·80	2·36
22	7·95	5·72	4·82	4·31	3·99	3·76	3·59	3·45	3·26	3·12	2·75	2·31
23	7·88	5·66	4·76	4·26	3·94	3·71	3·54	3·41	3·21	3·07	2·70	2·26
24	7·82	5·61	4·72	4·22	3·90	3·67	3·50	3·36	3·17	3·03	2·66	2·21
25	7·77	5·57	4·68	4·18	3·86	3·63	3·46	3·32	3·13	2·99	2·62	2·17
26	7·72	5·53	4·64	4·14	3·82	3·59	3·42	3·29	3·09	2·96	2·58	2·13
27	7·68	5·49	4·60	4·11	3·78	3·56	3·39	3·26	3·06	2·93	2·55	2·10
28	7·64	5·45	4·57	4·07	3·75	3·53	3·36	3·23	3·03	2·90	2·52	2·06
29	7·60	5·42	4·54	4·04	3·73	3·50	3·33	3·20	3·00	2·87	2·49	2·03
30	7·56	5·39	4·51	4·02	3·70	3·47	3·30	3·17	2·98	2·84	2·47	2·01
32	7·50	5·34	4·46	3·97	3·65	3·43	3·26	3·13	2·93	2·80	2·42	1·96
34	7·45	5·29	4·42	3·93	3·61	3·39	3·22	3·09	2·90	2·76	2·38	1·91
36	7·40	5·25	4·38	3·89	3·58	3·35	3·18	3·05	2·86	2·72	2·35	1·87
38	7·35	5·21	4·34	3·86	3·54	3·32	3·15	3·02	2·83	2·69	2·32	1·84
40	7·31	5·18	4·31	3·83	3·51	3·29	3·12	2·99	2·80	2·66	2·29	1·80
60	7·08	4·98	4·13	3·65	3·34	3·12	2·95	2·82	2·63	2·50	2·12	1·60
120	6·85	4·79	3·95	3·48	3·17	2·96	2·79	2·66	2·47	2·34	1·95	1·38
∞	6·63	4·61	3·78	3·32	3·02	2·80	2·64	2·51	2·32	2·18	1·79	1·00

Table 7   *Conversion of Range to Standard Deviation*

n	a_n	n	a_n	n	a_n	n	a_n
2	0·8862	5	0·4299	8	0·3512	11	0·3152
3	0·5908	6	0·3946	9	0·3367	12	0·3069
4	0·4857	7	0·3698	10	0·3249	13	0·2998

An estimate of the standard deviation is given by multiplying the range of a random sample of size n from a normal population, by a_n. The mean range in samples of size n from a normal population is the standard deviation of the population divided by a_n.

Table 8 Distribution of Spearman's rank correlation coefficient in random rankings. Probability P that a given value of Σd^2 will be attained or exceeded

$n=4$		$n=5$		$n=6$		$n=7$		$n=8$		$n=9$		$n=10$	
Σd^2	P	Σd^2	P	Σd^2	P	Σd^2	P	Σd^2	P	Σd^2	P	Σd^2	P
12	0·458	22	0·475	50	0·210	74	0·249	108	0·250	156	0·218	208	0·235
14	0·375	24	0·392	52	0·178	78	0·198	114	0·195	164	0·168	218	0·184
16	0·208	26	0·342	54	0·149	82	0·151	120	0·150	172	0·125	228	0·139
18	0·167	28	0·258	56	0·121	86	0·118	126	0·108	180	0·089	238	0·102
20	0·042	30	0·225	58	0·088	90	0·083	132	0·076	188	0·060	248	0·072
		32	0·175	60	0·068	94	0·055	138	0·048	196	0·038	258	0·048
		34	0·117	62	0·051	98	0·033	144	0·029	204	0·022	268	0·030
		36	0·067	64	0·029	102	0·017	150	0·014	212	0·011	278	0·017
		38	0·042	66	0·017	106	0·0062	156	0·0054	220	0·0041	288	0·0087
		40	0·0083	68	0·0083	110	0·0014	162	0·0011	228	0·0010	298	0·0036
				70	0·0014							308	0·0011
20		40		70		112		168		240		330	

Tail area at the lower end of the distribution.

Σd^2 is symmetrically distributed about $\frac{1}{3}(n^3-n)$. P = the probability that Σd^2 is equalled or exceeded. Since the upper and lower limits for Σd^2 are 0 and $\frac{2}{3}(n^3-n)$, respectively, the probability is also P that the sum of squares of the rank differences is less than or equal to $\frac{1}{3}(n^3-n)-\Sigma d^2$. To facilitate computation of the lower tail significance levels of the distribution of Σd^2, values of $\frac{1}{3}(n^3-n)-\Sigma d^2$ are given in the bottom row of the table. For example, if $n=8$, the probability is 0·048 that the sum of squares of the rank differences will be less than or equal to $168-138=30$.

Source: Pearson and Hartley (ed.), Biometrika Tables for Statisticians, Vol. 1.

Index

347